SpringerBriefs in Space Development

Series Editor: Joseph N. Pelton, Jr.

For further volumes:
http://www.springer.com/series/10058

Don M. Flournoy

Solar Power Satellites

Don M. Flournoy, Ph.D.
Professor of Telecommunications
Scripps College of Communication
Ohio Center of Excellence
Ohio University
Athens, OH 45701, USA
don.flournoy@ohio.edu

ISBN 978-1-4614-1999-0   e-ISBN 978-1-4614-2000-2
DOI 10.1007/978-1-4614-2000-2
Springer New York Dordrecht Heidelberg London

Library of Congress Control Number: 2011943353

Printed on acid-free paper

Springer is part of Springer Science+Business Media (www.springer.com)

This Springer book is published in collaboration with the International Space University. At its central campus in Strasbourg, France, and at various locations around the world, the ISU provides graduate-level training to the future leaders of the global space community. The university offers a two-month Space Studies Program, a five-week Southern Hemisphere Program, a one-year Executive MBA and a one-year Masters program related to space science, space engineering, systems engineering, space policy and law, business and management, and space and society.

These programs give international graduate students and young space professionals the opportunity to learn while solving complex problems in an intercultural environment. Since its founding in 1987, the International Space University has graduated more than 3,000 students from 100 countries, creating an international network of professionals and leaders. ISU faculty and lecturers from around the world have published hundreds of books and articles on space exploration, applications, science and development.

Preface

Don't be intimidated by this topic. Although collecting energy from space might appear to be a distant and overly complex answer to our energy problems on Earth, a careful look reveals surprising advantages. This small book is intended to be a "quick study" overview of space solar power, its current status and prospects for implementation, a topic that could matter to you and me personally.

You may find that solar power from space is not so far off. As one who has edited a journal on space communication and was a board member of the Society of Satellite Professionals International for a decade, it is increasingly clear to me that our next generation of space satellites will be of the power satellite–Sunsat–variety. Solar power satellites will be launched for the principal purpose of capturing the Sun's energy in space and delivering it to Earth as a non-polluting form of electrical power. These new Sunsats, I predict, will not only serve as the basis for the revitalization of the space industry, they will be a key to the future economic strength and environmental health of all nations.

Solar Power Satellites makes the case that space solar power is poised to become the planet's most significant source of alternative (clean and renewable) energy. True, space satellites will not soon replace the infrastructure and business models for terrestrial energy production and distribution, but the Sunsat systems promise to be a complementary source of global "base load power," i.e., electrical power that can be accessed and delivered whenever and wherever it might be needed.

We know that energy demand is growing. We also know that all current sources of energy will sooner or later prove to be insufficient, either due to declining production, as with oil and gas, or environmental concerns, as with coal or nuclear, or the insufficiencies of terrestrial wind and solar. How will our future needs be met? In this book, I make the case that to bring "power to all the people everywhere," only a space-based global power grid will do the job. To make such a vision a reality, individual governments will certainly have to play a supportive role – to assure that their economies continue to grow and their citizens have a reasonable quality of life – but to develop an energy market of such size and scope commercial involvement is required.

In brief, the task will require: 1) larger and more sophisticated space platforms, arrays and power transmission systems; 2) more robust and reliable transportation systems for delivering materials to space; 3) specialized large-scale receivers, converters, storage and distribution systems on Earth; 4) in-orbit position allocations and assignment of radio frequency spectrum for energy transmission; and 5) effective operational arrangements and management systems to insure that all components work together efficiently and safely.

Accomplishing these goals will obviously require financial, intellectual and diplomatic resources in considerable portions. But being successful will ultimately mean a lot to Planet Earth, including economic renewal of our moribund communities, the creation of new businesses and jobs, cleaner air and water, more stable weather and climate and possible avoidance of energy-related conflicts.

Solar Power Satellites is unique for its coverage of three emerging situations: 1) the social and economic pressures everywhere that require new energy solutions; 2) the growing recognition that space-based solar power is an unlimited, non-polluting source of new energy; and 3) the financial and business opportunities that will attract the aerospace, communication satellite and related industries to this new market.

The book is written for the non-technical professional and interested student. Illustrative examples are drawn from the space industry, from energy sectors, and from basic science. Explanations are straightforward; the language is easy to follow and understand. I believe that scientists, engineers, economists, and regulatory authorities will also find the overview and evidence presented to be timely and informational.

Athens, OH, USA   Don M. Flournoy, Ph.D.

Contents

1 What Is a Solar Power Satellite? 1
  What Is a Sunsat? 1
  Power Plants in Space 2
    The Space Segment 3
    The Launch Segment 4
    The Ground Segment 5
  Challenges That Sunsats Face 6
  A Perfect Storm 7
  Concluding Thoughts 8
  References 8

2 What Are the Principal Sunsat Services and Markets? 9
  The Energy Picture 9
  Climate Change 10
  Satellite Power Markets 11
    Power-to-Power Utilities 12
    Power-to-Agriculture 12
    Power-to-Terrestrial Solar 13
    Power-to-Fresh Water 14
  Power-to-Cities 14
  Power-to-Disaster Sites 16
  Concluding Thoughts 17
  References 17

3 What Will Sunsats Look Like? 19
  Technical Feasibility 19
  Commercial Viability 20
  New Architectures 21
  Newer Research 23
  Other Technical Challenges 24

Corporate Research 25
Concluding Thoughts 26
References 26

4 **How Will Sunsats Be Delivered to Space?** 29
Launching Sunsats 29
An Historical Perspective 30
Launch Strategies 31
Reducing Costs 33
Reusable Rockets 35
Alternative Approaches 36
Concluding Thoughts 37
References 38

5 **How Will Sunsat Power Be Captured on Earth?** 39
Future Prospects 39
Historical Perspective 41
Public Policy Concerns 42
  Environment and Health 42
  Upper Atmosphere Effects 43
  Land Use 44
  Space Communications 45
Concluding Thoughts 45
References 46

6 **What Is the Economic Basis for Solar Power Satellites?** 47
The Case for Sunsats 47
Bilateral Project Development 48
Indo-U.S. Collaboration 50
The Commercial Sector 51
Intermediate Steps 52
Concluding Thoughts 53
References 54

7 **What Are the Legal Issues?** 55
International Development Goals 55
Space Law 57
  The Outer Space Treaty 57
  The Liability Convention 58
  The Registration Convention 59
  Space Debris 60
  Microwave Radiation 62
Other Regulatory Issues 62
  GEO Slot Rights 62
  Power Beaming 63
  Renewable Energy Targets 64
The Role of Government 65
Concluding Thoughts 65
References 66

8   How Is Sunsat Development Faring Internationally?   67
    SPS over China   67
        China's Long-Term Vision..   69
        China's Energy Future   69
        Sustainable Development...   70
        A Skilled Workforce   70
    Heading Off and Mitigating Disasters   71
        SPS Implementation..   71
    SPS over India...   72
        India's Energy Policies..   73
        India's Strategic Goals   73
        Power Capacity Constraints   74
    SPS over Japan...   75
        The Japanese National Space Plan...   76
    International Cooperation and Collaboration..   77
    Concluding Thoughts..   77
    References..   78

9   What Is Worrisome About Solar Power Satellites?   79
    Launch to Space...   79
    Assembly in Space..   80
    Wireless Transfer of Energy   80
    Land Use   81
    Satellite Collisions   82
    Space Debris   83
    Solar Storms and Flares   83
    Signal Interference   84
    Concluding Thoughts..   84
    References..   85

10   Top Ten Things to Know About Space Solar Power   87
    References..   99

Glossary   101

About the Author   103

Index...   105

Chapter 1
What Is a Solar Power Satellite?

Abstract This introductory chapter explains how the new Sunsats—sometimes called powersats or solarsats—will differ from comsats in terms of purpose, operations, market, regulation and design.

What Is a Sunsat?

A solar power satellite is a space-based vehicle for gathering quantities of sunlight in space and delivering it to Earth as electrical power. Such satellites are poised to become the next-generation equivalent of communication satellites, and energy services will be their new market.

No solar power satellites are yet in operation. While all satellites in Earth orbit host some type of solar collector to generate the energy for power and control, no such satellites are there for the primary purpose of gathering energy from the Sun and delivering it to Earth. Because an abundant and sustainable new source of energy is desperately needed on Earth and the current level of technological development will now permit it, a huge new satellite sector is about to emerge that will relay energy from space to antennas on the ground, where it will be used on-site or plugged into our electrical power grids.

The logical path forward for those intending to develop solar power generation plants in space is in partnership with the commercial satellite (comsat) industry, a well established ($200 billion per year) sector with 40-plus years of expertise in designing, manufacturing, launching and operating spacecraft in orbit above Earth.

The future is never very clear, but once it becomes clear that communication satellites can be repurposed to safely and profitably deliver energy as well as video, voice and data signals, the author predicts it will be the comsat stakeholders taking the lead in new Sunsat ventures. This is logical; near-space is their home territory. They will enter into the field with the global perspective, the venture capital, the regulatory clout, the managerial experience and the marketing skills to turn such an enterprise into multiple viable businesses.

D.M. Flournoy, *Solar Power Satellites*, SpringerBriefs in Space Development, DOI 10.1007/978-1-4614-2000-2_1, © Don M. Flournoy 2012

Power Plants in Space

The idea that the Sun's rays can be collected in space and beamed to Earth as an energy source from a space-based platform has been around even longer than the idea of communications satellites. The entrepreneurs in communications sectors were the first to commit to space because they were quicker to see the advantages in having transmission towers located high above Earth for widest reach, coverage and mobility, while the power industry stayed Earthbound, feeling assured that it could meet future demand by scrapping for fossil fuels on the ground.

In the 1970s, Earth-based energy was still readily available and very cheap. But some 40 years later, oil and gas reserves are harder to find and a lot more expensive to retrieve. Coal and nuclear fission material are perceived as "dirty energy" sources. Also, by the twenty-first century, all the unaccounted for costs of environmental desecration and atmospheric pollution associated with fossil fuels have finally come due just when long-term energy security for many nations is in doubt.

Those scenarios, along with some prominent disasters in the energy business, have created the context for a more favorable reconsideration of the Sunsat option. Although initial investment costs are still considered high, the attractiveness of clean, abundant and instantly useful energy drawn down from strategically placed solar stations above Earth is now too compelling to ignore (Fig. 1.1).

Since comsats and Sunsats have many similar technological and operational requirements, it is worth considering how their business plans might converge. These in-orbit satellites perform a variety of functions, the most significant being communication (audio and video broadcasting, mobile telephony, broadband data and Internet); remote sensing (weather, environmental surveillance, mapping); and geo-positioning

Fig. 1.1 The Sun Tower is a conceptual design based on NASA's 1997 Fresh Look study in which the transmitter diameter is 500 m and the vertical "backbone" length is 15.3 km. An equally large rectenna receiver provides for power production on the ground (Potter et al. 2009)

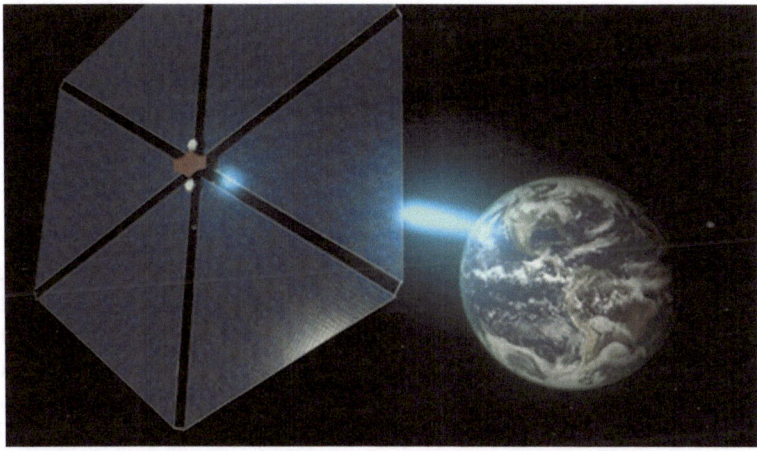

Fig. 1.2 Solar power satellite design created by Ohio University students affiliated with the Game Research and Immersive Design Laboratory (GRID Lab), commissioned by the Online Journal of Space Communication for the 2011–2014 Sunsat Design Competition (Ohio 2011)

and navigation. As a platform for performing work beyond Earth's atmosphere, the International Space Station (ISS) is also a multipurpose satellite, conducting research while testing the opportunities and challenges of living and working in space.

Will solar power satellites differ radically from those operating in space today? The answer is yes—and no. If one considers the three basic structural elements of communication satellites—that is, the space segment, the Earth segment and the transport segment, one can see that they have much in common.

The Space Segment

The new solar power satellite industry will position above Earth a new type of energy infrastructure hosting many of the features of communications platforms, including a satellite bus (physical structure), solar arrays, onboard processing, telemetry control and wireless transmission systems. Unlike the comsats that gather a small portion of the Sun's radiation to power their spacecraft, the Sunsat antennas would be designed to collect and concentrate solar thermal or photovoltaic energy for the principal purpose of relaying it to Earth, where it will be converted into electricity (Fig. 1.2).

While development of the thinner, lighter, cheaper photovoltaic (PV) cells that make terrestrial power production increasingly more efficient currently benefits communications systems in space, the benefits will be much greater for solar power producers looking to reduce the size and increase the productivity of their antennas while holding down the costs of launching their much larger solar collection arrays into space. Also benefitting the Sunsat and comsat industries will be promising new developments in remote construction, assembly, repair and replacement.

Sunsats will need bigger, more efficient solar panels than are currently in use since the principal purpose of their onboard power conversion and transmission systems will be to convert the Sun's energy into low-density radio or light frequency waves capable of providing many times more electrical power than we use today. To increase efficiency, large-scale reflectors will be used to concentrate photons from the Sun such in a way that the PV cells see the equivalent of not just one Sun but many suns.

Among the more innovative Sunsat designs are architectures that network more than one satellite together within a common space orbit, creating a photovoltaic area of 20 km or more. Multiple clusters of such satellites may one day be operating in space orbit, and these will be linked for global electric power service. While building, launching and assembling such structures in space will be a massive undertaking, past space achievements (such as the International Space Station, the Hubble Telescope, the Mars rovers and the many spacecraft that operate safely and productively in Earth orbit) give proponents of space solar power increased confidence that locating solar stations in space is within our reach.

Comsat architectures in the digital age have greatly improved functionality and performance as a result of onboard computer processing and control, and effective use of spot beam technologies. These advanced technology spacecraft can direct communications transmissions to more narrowly defined regions and increase power levels through cloud cover. Such beams can be moved from one receiver to another on command from Earth. While transmitting a communication signal requires significantly different operations from those required in wireless power transmission, these more advanced comsat designs will help to solve some of the challenges faced by Sunsat engineers.

The Launch Segment

Launch systems are key to space-based solar power implementation. Every piece of infrastructure destined for space must be shoved out of Earth's gravitational field using one or more of the principal launch vehicle types. These include a wide variety of reusable launch vehicles (RLV). Some are of these are of the "vertical takeoff vertical landing" and "horizontal takeoff horizontal landing" types; some are "single stage to orbit" or "two-stage to orbit," with the first stage from the ground. Other options are in development (Bienhoff 2008, p. 2).

At least in the beginning, Sunsats will employ the same private, commercial and government rockets used to lift communications satellite structures from Earth to space. Some plans involve assembling solar satellites and their antennas from components lifted by medium power rockets into a low-Earth orbit (LEO), possibly using the International Space Station as a staging area, later transferring the assembled unit into its final position in a geosynchronous, Sun-synchronous or other suitable orbit. Other plans call for inserting solar spacecraft and their large arrays directly into the designated orbit using more powerful thrusters.

Launching satellites safely and economically into space is among the greatest challenges of the satellite industry. But after many years of successes and failures, the industry is consistently delivering 90% of its payloads into designated orbits. This level of predictability will give the energy providers, as well as the insurance business, a high level of confidence that the launch providers can do what they say they can do.

The communications industry is now—and the solar power industry will soon be—the beneficiary of an ongoing global effort to regularize space transport, making it a viable business enterprise in the way that aerospace is today. To avoid the high costs of launching workers and material into space, some visionaries see space-based infrastructures being built from materials found on the Moon (and on near-Earth asteroids), with robotic manufacturing and assembly managed from Earth via virtual systems of communications and control. The orbits above the Van Allen radiation belts, where the Sunsats will operate, are too intense as a radiation environment for long-term workers, so most Sunsat construction and maintenance is expected to be done tele-robotically—by operators on the ground.

The Ground Segment

Rectifying antennas—Earth receiving stations—will capture the transmitted signals of the solar satellites and convert them into electrical power. In this respect, Sunsat receivers will resemble the passive early TVRO (receive only) Earth antennas of radio and television, capturing not information but energy to be relayed to clients and consumers. Except for telemetry (and the low power guide beam originating with the ground receiver that insures the satellite transmitter is focusing its main power beam accurately), no uplink is needed. The Sunsat on-ground receivers will also be substantially larger than those of radio/TV, lowering the energy density to acceptable levels.

Were the power levels to be too focused, there could be dangerous effects. Highly concentrated transmissions from space could harm airline passengers flying through the RF (radio frequency) beam. Reflections from the reception antennas could interfere with or disable the communications of other application satellites. The answer is to create low-density RF energy beams and spread them more broadly. With networked arrays capable of producing electrical rating equivalents of coal fired or nuclear power plants at 1 gw or larger, solar power rectennas can be expected to stretch 1–10 kilometers (km) across. Such collection points will require a protected area similar to that established with coal and nuclear plants; their advantage, however, is that agricultural crops can be grown and fish farms and greenhouses can be situated on Sunsat sites. The fuel they use will not have been extracted from Earth; the power they will generate will be non-polluting and there will be no toxic waste to be disposed of.

Just as satellite communication teleports and antenna farms are connected into the broadband fiber optic networks distributing signals terrestrially, Sunsat antenna farms will be connected into a terrestrial grid that distributes electrical power. While comsats are networked with data centers for information storage and retrieval, the Sunsats will be networked into power distribution centers that will ensure balanced energy transmissions within regions served by their multiple electricity sources.

Challenges That Sunsats Face

As with communication satellites, solar power satellites must be lifted into designated orbits, where they will be expected to provide service to specified regions. No matter the orbit, such satellites must go through a nation-by-nation approval process that will ultimately involve the International Telecommunications Union, an agency of the United Nations, to decide upon a particular location and type of service.

World satellite communications is strictly regulated in terms of orbital registration and position, frequency allocations and levels of power transmission. Since the solar power satellite industry will be arriving late in the process, it will encounter some resistance on such matters as orbital slots and frequency assignments, as these are by nature scarce. The commonly discussed orbital location for Sunsat placement is the geosynchronous Earth orbit, a 36,000-km-high "sweet spot" heavily used by communications satellite services.

Minimizing interference with electromagnetic spectrum assignments of other space users and with those on the ground is the principal reason for such controls. Although none of the existing players in space will be conducting businesses in direct competition with Sunsat products and services, some resistance to sharing positions and spectrum is to be expected from incumbents protecting performance (and future) of their communications, navigation, remote sensing and other systems.

In January 2008, the Space Solar Power Institute of Atlanta, Ga., approached the U.S. House Committee on Science and Technology with a proposal to form "a congressionally chartered public/private corporation" patterned after the highly successful model provided by the COMSAT Act of 1962. That model led to the creation of the Intelsat (international satellite) consortium that now provides satellite communication to all world regions (Preble 2008). The purpose of the proposed Sun Satellite Corporation would be "to build commercial power satellites to collect and transmit energy to electric power grids under contract to wholesale (utility) customers on Earth." The strategy was offered as a concrete step forward in improving U.S. energy security.

News about this initiative made few headlines and has all but disappeared from view, but the idea of private/public corporations focusing on new energy resources is very much alive. It now appears that development of space solar power in the United States will be a lot more private sector than government driven. Private/public collaborations are also the most likely approaches to be taken by such space-faring nations as Canada, China, India, Japan, Russia and the European Union, where solar power satellite systems will be launched by collaborators as often as by competitors in the race to space for energy. The idea of Sunsat corporations is not going away in the United States or elsewhere.

The most significant barriers to realizing a new satellite business based on energy from space are not technological. Certainly there are many technical challenges to be met. These include easier and cheaper access to space, greater efficiencies and capacities of solar cells, wireless power transmission and receiver networks, and energy conversion, storage and distribution systems.

Space visionaries have always looked to governments to get ambitious projects off the ground. In the building of Sunsat infrastructures, governments can help with research and development funding, assist with demonstration projects and agree to be the anchor tenant purchasing the first products produced. But today, countries around the world are expecting their commercial sectors to be involved, and involved early, for creative design as well as for long-run implementation and management.

Progress in raising capital for Sunsat businesses will inevitably be tied to progress in space commercialization overall, and the development of plausible business plans related to alternative energy markets in particular. The fact that the U.S. demand for electricity is expected to increase by as much as 40% in the next two decades, and assumptions that lesser developed nations will wish to grow even faster, is a key incentive. The rising cost of conventional carbon-based energy sources, coupled with the increasing cost of overcoming greenhouse gas pollution and safety concerns associated with nuclear energy, are helping to move up the timetable.

A Perfect Storm

The world is facing a perfect storm in which an energy crisis and an environmental crisis are occurring simultaneously. Earth's population continues to grow. Oil, gas and coal, the principal energy basis for the steadily improving standards of living among the more developed societies—and coveted by lesser developed societies— are now shown to be contaminating Earth's atmosphere. Atmospheric pollution and climate change occur as carbon-based fuels are mined, processed and consumed. At the same time, those nonrenewable fossil fuels are rapidly being used up. Experts predict that, within the next generation, fossil fuels—plus all known alternative energy sources on Earth—will fall far short of projected need.

Several government commissions, think tanks, energy companies and utilities in more than one country investigating the potential of space-based solar power have concluded that satellite delivery must be a part of the long-term solution. Such studies note that the solar energy available in space is several billion times greater than any amount human societies could ever use on Earth. The Sun's energy, always available, is virtually inexhaustible. Unlike fossil fuels, space solar power does not emit greenhouse gases. Moving to solar energy can also reduce competition for the limited supplies of Earth-based energy, predicted to be the basis for future wars.

Prior to its 2011 nuclear disaster, Japan had already made a financial commitment to go into space for one of its long-term alternative energy solutions. In September 2009, a research group representing 16 companies, including Mitsubishi Electric and Mitsubishi Heavy Industries, announced a 2 trillion yen ($21 billion) effort to build and launch into GEO a 1 gw solar station, to be in operation by 2025 (Sato and Okada 2009). As proposed, the satellite was to be fitted with 4 km² of solar panels. In 2015, a smaller demonstration satellite fitted with wireless power transmission equipment was to be used to test power beaming to Earth (Yomiuri Shimbun 2011). Since its March 2011 nuclear disaster, Japan's resolve to build SSP has apparently escalated, although not scheduled to become available until significantly more R & D is done.

Solaren, a U.S.-based entrepreneurial company, has indicated its plan to deploy an alternative design on a more accelerated schedule. This innovative design consists of several components. One is a series of concentrator reflectors that would focus power from the Sun so that the solar array would see the equivalent of many suns. The second is a solar array that would be of higher efficiency and have longer life that would convert the solar energy into power. The third is a transmission system that would relay the energy to Earth as RF (microwave) power. This company has signed contracts to deliver energy to U.S. West Coast public utilities starting as early as 2016. These contracts, however, are non-binding and go into effect only when Solaren is actually able to start delivering space solar power at a commercially viable rate (Bullis 2009).

Concluding Thoughts

Figuring out how to collect energy in space and transmit it on demand to anywhere on Earth will be an undertaking of far greater significance than placing a man on the Moon or building a human habitat on Mars. Such an accomplishment—ready access to energy on Earth (and elsewhere)—is key to all space exploration. Because Sunsats can tap the one energy supply that cannot be depleted, any corporation or country in the space energy business will have a perpetual competitive advantage.

In practical terms, building international businesses around solar energy from space may be the only way we can keep alive our individual and collective dreams for a better life. Having abundant, safe, non-polluting energy could represent a tipping point for human productivity and creativity—that one essential ingredient enabling the human race not just to survive but to live up to its potential. If indeed solar energy can make that difference, let us work toward the possibility, as there are no other sustainable solutions currently available to meet our seemingly unending demands for power.

References

Bienhoff, D. 2008. Space infrastructure options for space based solar power. 6th International Energy Conversion Engineering Conference (IECEC). 28–30 July 2008. Cleveland, Ohio.
Bullis, K. 2009. Startup to beam power from space. *Technology Review*, April 15. http://www.technologyreview.com/blog/energy/23381/. Accessed 5 July 2011.
Ohio University GRID Lab. 2011. http://sunsat.gridlab.ohio.edu.
Potter, S., M. Bayer, D. Davis, A. Born, D. McCormick, L. Dorazio, & P. Patel. Space solar power satellite alternatives and architectures. AIAA Aerospace Sciences Meeting. Orlando, Florida, 5–8 January 2009.
Preble, D. 2008. How to build a space solar power system: The Sunsat Incorporation Act. Space Solar Power Workshop. http://www.sspi.gatech.edu. Accessed 23 January 2008.
Sato, S. & Y. Okada. 2009. Mitsubishi, IHI to join $21 billion space solar project. *Bloomberg*. http://www.bloomberg.com. Accessed 15 September 2009.
Yomiuri Shimbun. 2011. Space-based solar power set for 1st test. http://www.yomiuri.co.jp/ 23 January 2011.

Chapter 2
What Are the Principal Sunsat Services and Markets?

Abstract This chapter describes some of the challenges facing the planet as a result of burning fossil fuels, and the opportunities presented to the satellite industry in response to world demand for cleaner and more abundant energy. Among the Sunsat uses discussed are the production of baseload electrical power—not just an intermittent source of power—supporting agriculture, saltwater desalination, disaster relief, military operations and related applications.

The Energy Picture

Gordon Woodcock, a space scientist for the Boeing Company who has worked on SPS solutions for more than 30 years, defines fossil fuels as "solar energy stored in chemical form by natural processes over hundreds of millions of years." He observes that we are depleting this stored energy in a timeframe measured in decades rather than millennia. Present methods of fossil fuel consumption are also extremely dirty, polluting our Earth's biosphere. It's a race between resource depletion and destruction of the environment. Either way, he says, the world economy collapses (Woodcock 2010).

Alternative terrestrial energy is not the complete answer, either. According to Woodcock, the limitation of Earth-based renewable energy sources is that they are not "demand" sources; that is, they are only intermittently available. Terrestrial solar power works when the Sun shines. Terrestrial wind power works when the wind blows.

Terrestrial hydroelectric power is a way of storing water energy until users demand it. This process can include hydroelectric pumped storage, which is the lifting of water uphill where it is held until released to create electricity as it flows through turbines. But there is little capacity remaining on the planet for hydroelectric installations. Geothermal energy is also way to tap stored energy in the Earth itself. Batteries, water electrolysis and hydrogen storage in fuel cells are other ways to provide storage. But to run a modern power grid exclusively (or even largely) on terrestrial renewable energy, he says, would require enormous amounts of storage, and storage is expensive.

D.M. Flournoy, *Solar Power Satellites*, SpringerBriefs in Space Development, DOI 10.1007/978-1-4614-2000-2_2, © Don M. Flournoy 2012

Woodcock concludes that solar power satellites are a potential solution because they can be positioned in space over a particular location to which they can stream continuous sunlight. Supplying power around the clock, such an energy system can serve as a demand source with very little storage required. He also suggests, given constant solar pointing, the photovoltaic area could probably be reduced by a factor of 10–100 by using concentrators. Land designated for receiving sites might also serve dual or multiple purposes.

The National Space Society (NSS) hosts annual conferences that include sessions on space solar power. The organization's website includes one of the most complete archives on space solar research. It also has taken positions of advocacy, stating that "all viable energy options should be pursued with vigor, [but that] Sun/Sat power (SSP) has a number of substantial advantages over other energy sources." The NSS lists several of these advantages:

- Unlike oil, gas, ethanol and coal, SSP does not emit greenhouse gases.
- Unlike nuclear power plants, SSP does not produce hazardous waste that needs to be stored and guarded for hundreds of years.
- Unlike terrestrial solar and wind power plants, SSP can be available in huge quantities 24-hours-a-day, 7 days a week. It produces regardless of cloud cover, daylight, or wind speed.
- Unlike coal and nuclear fuels, SSP does not require environmentally problematic mining operations.
- Unlike nuclear power plants, SSP does not provide potential targets for terrorists (National Space Society 2008).

The National Space Society notes that "SSP will provide true energy independence for the nations that develop it, thereby eliminating a major source of national competition for limited Earth-based energy resources." The society acknowledges that "SSP development costs will be large, although significantly smaller than that of the American military presence in the Persian Gulf or those associated with the impacts of global warming" (National Space Society 2008).

Climate Change

In the *Online Journal of Space Communication*, Dr. Feng Hsu, a NASA scientist at Goddard Space Flight Center, a research center in the forefront of science of space and Earth, writes, "The evidence of global warming is alarming," noting the potential for a catastrophic planetary climate change is real and troubling (Hsu 2010).

Hsu and his NASA colleagues were engaged in monitoring and analyzing climate changes on a global scale, through which they received first-hand scientific information and data relating to global warming issues, including the dynamics of polar ice cap melting. After discussing this research with colleagues who were world experts on the subject, he wrote:

I now have no doubt global temperatures are rising, and that global warming is a serious problem confronting all of humanity. No matter whether these trends are due to human

Fig. 2.1 As an indicator of global warming, the glaciers on the South Island of New Zealand are observed melting (Photo by the author)

interference or to the cosmic cycling of our solar system, there are two basic facts that are crystal clear: (a) there is overwhelming scientific evidence showing positive correlations between the level of CO_2 concentrations in Earth's atmosphere with respect to the historical fluctuations of global temperature changes; and (b) the overwhelming majority of the world's scientific community is in agreement about the risks of a potential catastrophic global climate change. That is, if we humans continue to ignore this problem and do nothing, if we continue dumping huge quantities of greenhouse gases into Earth's biosphere, humanity will be at dire risk (Hsu 2010).

As a technology risk assessment expert, Hsu says he can show with some confidence that the planet will face more risk doing nothing to curb its fossil-based energy addictions than it will in making a fundamental shift in its energy supply. "This," he writes, "is because the risks of a catastrophic anthropogenic climate change can be potentially the extinction of human species, a risk that is simply too high for us to take any chances" (Hsu 2010).

It was this NASA scientist's conclusion that humankind must now embark on the next era of "sustainable energy consumption and re-supply, the most obvious source of which is the mighty energy resource of our Sun" (Hsu 2010) (Fig. 2.1).

Satellite Power Markets

This new energy market will have lots of stakeholders. Those who contribute to the energy supply and those who receive benefits from an on-demand power resource will represent all sectors in all nations, including business and commerce, government and military and the public at large.

Energy is a $1 trillion-plus global industry, and demand is expected to double every 20 years (Mankins 1997, p. 8), making this market attractive enough to bring it to the attention of the global satellite industry. From their inception, space satellites have collected and used the Sun's rays as a power source for communication and related services. Were satellite services to extend their range of offerings to include energy production, baseload electrical power (and other applications) could conceivably become a major new product line. Here are some illustrative examples.

Power-to-Power Utilities

One of the obvious opportunities for solar power satellites is to become an on-demand source of electric power for terrestrial utilities. Once Sunsat providers can demonstrate the capability to direct continuous radio or light frequency power beams to production sites, the owners of coal-fired generation stations will quickly discover the value of this service. The same will also be true of nuclear, gas-fired, biomass and other such plants.

With electrical power production ratings of 1 gw or more, solar satellite systems can be designed to meet the short- and long-term needs of the terrestrial power plants at their existing locations, at first to complement but eventually to replace their current fuel feedstocks. An attractive feature of this approach for space solar power investors is that the utilities have a predictable need for energy in great quantities. Since the power utilities are already connected to an electrical power grid, often covering regions larger than a single state or nation, the Sunsat people won't have to also be in the terrestrial distribution business.

Whether producing power from coal, nuclear, gas, biomass or other sources, power utilities can be expected to step forward as early users of this new space asset to begin reducing their mining and transportation costs. The use of scrubbers and filters will be greatly reduced, if needed at all. Problems related to spent fuel disposal and toxic waste management should be fewer. But mainly the utilities will become clients (and possibly investors) in the Sunsat business to guarantee a sustainable night-and-day fuel source.

Power-to-Agriculture

In many places on Earth, the climate, soil and terrain does not permit cultivation. With innovative applications of space solar power, it may be possible to establish multipurpose greenhouses and other agricultural facilities above which space-pointing Earth antennas have been installed for the purposes of producing heat along with electricity.

An example is reclaimed strip mine land brought back to productive use with the cultivation of local vegetables, flowers and other high-value crops underneath a several kilometer space solar power antenna. In this scenario, the SPS rectenna is a wire mesh energy receiver positioned above the greenhouses. The constant temperatures and light created in the generation of energy make for a 12-month growing season. The wire mesh energy receiver produces electricity that can be used to operate machinery and supply the local power grid. This approach creates a business circle: an environmentally friendly energy production operation that can take advantage of seemingly worthless land to produce cash crops and have access to readily usable energy to stimulate the creation of new businesses, thereby improving the rural economy.

Power-to-Terrestrial Solar

A slight modification of the power-to-agriculture approach will be the design and installation of an SPS rectenna that covers a terrestrial solar generation site, as in the case of solar farms. Energy beaming from space would be coordinated to operate in sync with photovoltaic stations on the ground where Earth solar and space solar antennas are co-located, taking maximum advantage of the sunlight that makes its way through Earth's atmospheric filter and also using the microwaved energy, beamed from space, on the same unit.

Engineers have already figured out that photovoltaic arrays can be designed with an integral antenna built-in, thereby maximizing efficiency, or such systems can be constructed with the space solar collectors working overhead. In such cases, the dual-use installation assures 24-h power production (Landis 2004).

The Boeing Company sponsored space solar power research that looked at the matter of "synergy with other energy sources." In addition to finding that SSP required lower land use per unit power compared to other renewables, the team learned that "space solar power microwave antennas can be designed to let light pass through, so the same land area can be used for conventional solar power—or possibly agriculture" (Potter et al. 2009, p. 36).

Such installations do not yet exist, but the technical design and business plan for one of these could easily be modeled upon a project in Appalachian Ohio, where some 500 acres of reclaimed land, mined by the Central Ohio Coal Co. between 1969 and 1991, is expected to become the site of the largest solar farm in the eastern part of the United States.

Turning Point Solar's 49.9 mw solar array is to be built adjacent to The Wilds nature conservancy in Muskingum County, Ohio. In October 2010, American Electric Power signed a memorandum of understanding with project developers to enter into a 20-year purchase agreement for the facility's power. This project is aided by a 2008 energy reform bill that calls for 25% of all energy consumed by Ohioans to come from advanced energy sources by 2025 (Athens Messenger 2010).

Power-to-Fresh Water

One resource that has been negatively affected by the increasing accumulation of carbon dioxide and methane greenhouse gases in Earth's lower atmosphere is clean water. Kent Tobiska, a space environment scientist, says that one effect of adverse climate change is flooding and fresh water contamination. Population growth has also reduced water supplies while increasing demand.

Tobiska, in a paper written for the American Institute of Aeronautics and Astronautics (AAIA), notes that continued population growth in coastal areas makes it economically feasible to begin considering seawater desalination as a larger source for metropolitan water supplies. He also notes that the process of desalination is, however, energy intensive, which has discouraged its widespread use. (Tobiska 2009, p. 1)

In a later paper, written for the *Online Journal of Space Communication*, he made public an unusual proposition and proposal to the State of California, one that could help not only solve the state's energy problems but also allow coastal areas access to a continuous supply of fresh water. He writes:

> California offshore oil and gas platforms already use seawater desalination to produce fresh water for platform personnel and equipment. It is proposed that as California coastal oil and gas platforms come to the end of their productive lives, they be re-commissioned for use as large-scale fresh water production facilities.
>
> Solar arrays, mounted on the platforms, are able to provide some of the power needed for seawater desalination during the daytime. However, for efficient fresh water production, a facility must be operated 24 h a day. The use of solar power transmitted from orbiting satellites (Solar Power Satellites—SPS) to substantially augment the solar array power generated from natural sunlight is a feasible concept.
>
> The architecture of using an SPS in geosynchronous orbit (will) enable 24 h a day operations for fresh water production through seawater desalination. Production of industrial quantities of fresh water on re-commissioned oil and gas platforms, using energy transmitted from solar power satellites, is a breakthrough concept for addressing the pressing climate, water, and economic issues of the twenty-first century using space assets (Tobiska 2011) (Fig. 2.2).

Power-to-Cities

It is predicted that by 2020 there will be 26 mega-cities—defined as a population area of ten million or more—in the world, primarily in the newly industrialized third world (Landis 2004, p. 16). Almost all of these high population areas will be scrambling to find the energy resources to meet even basic needs, with the more prosperous cities already having teams of planners trying to find answers.

Here again, California can be used as illustration. In December 2009, the California Public Utilities Commission unanimously approved a power purchase agreement that its utility Pacific Gas and Electric (PG&E) had negotiated with a space solar power provider. The 15-year contract with Solaren Corp., a Manhattan

Fig. 2.2 The production of fresh water using space power is illustrated in this visualization created by Ohio University GRID Lab students (Ohio 2010). For animation and related content see http://spacejournal.ohio.edu/issue17/present.html

and California-based company, set the goal to begin beaming from space 1.7 gw of electricity by 2016 to a receiver antenna in Fresno.

In 2010, Solaren was looking to raise more than $100 million to develop its orbiting solar farm in space. The project would require billions of dollars, including rockets that are likely to cost $150 million each (Wang 2009).

As part of PG&E's commitment to providing more renewable energy to its customers, the utility was supporting a wide range of technologies, including wind, geothermal, biomass, wave and tidal, and at least a half dozen types of solar thermal and photovoltaic power. With the Solaren agreement, PG&E extended that approach to renewable energy from space. A PG&E in-house report states that, while the concept of space solar power makes sense, making it all work at an affordable cost is a major challenge, which Solaren says it can solve. Solaren's team includes satellite engineers and scientists, primarily from the U.S. Air Force and Hughes Aircraft Company, with decades of experience in the space industry. Its CEO, Gary Spirnak, was a spacecraft project engineer in the U.S. Air Force and director of advanced digital applications at Boeing Satellite Systems, among other positions (Marshall 2009).

Among the arguments given was that the energy available in space is eight to ten times greater than on Earth. There's no atmospheric or cloud interference, no loss of Sun at night, and no seasons, which means that delivered energy from space is a continuous baseload resource, not an intermittent source of power. Even if hard to reach, real estate in space is still free. Solaren would need to acquire land only for the receiving station, which it can locate near existing transmission lines. Where the rectenna is located can make some difference in reducing delays.

Power-to-Disaster Sites

SSP disaster relief was one of the visualizations created for the fall 2011 issue of the *Online Journal of Space Communication* by students affiliated with the Ohio University GRID (Game Research and Immersive Design) Lab. The technology imaging and animation project was also used in the journal's launch of its 2011–2014 SunSat Design Competition, sponsored by the Society of Satellite Professionals International and the National Space Society.

The concept for a future-oriented solar power satellite solution to disaster recovery came from the team mentor Dean E. Davis, aerospace systems engineer, Lockheed Martin Corporation (Davis 2011). From the team's technical brief, the following explanations were given, prefaced by this: "When a disaster strikes an area, rescue teams fly in from all over to give aid. But destruction to the local infrastructure greatly slows rescue efforts, wasting precious recovery time. Finding ways to quickly recover from power outages and to restore communications in large-scale disasters can help to ameliorate its devastating results. This technical brief explains some of the problems encountered in a disaster relief effort and illustrates how space solar power might help in the recovery" (Power 2011).

Illumination: In the context of natural or man-made disasters, rescue workers need to be able to work around the clock. Due to the absence of lighting, they are often limited to working full force only during the day. The lack of illumination can be addressed, in part, by satellites orbiting Earth. Networked in constellations, specially designed satellites will act as mirrors to reflect sunlight upon the spot facing a disaster situation. Each of these satellites will host a 100-m-thin film solar-reflecting mirror orbiting in a Sun-synchronous orbit. This orbit will be 600 km above Earth, inclining 98°. Potentially, these satellites could focus between 10,000 and 20,000 lumens of light, or about as much light as the Sun gives off in the daytime. This space-based asset will enable rescue workers to continue working at nighttime, thus making it possible to save time and lives.

Power: Light alone will not be sufficient, as areas struck by disaster will also likely need electrical power. Terrestrial power can be replaced by space solar power. Although the first constellation of Sun-synchronous (SEO) Earth-orbiting satellites provides light, imagine a second set of orbiting satellites. These satellites will convert the Sun's energy into electricity and beam it to Earth via laser-focused light beams operating at safe IR (infrared) or microwave frequencies. With giant solar collectors onboard, the satellites will collect energy via their solar cells and convert the energy into electrical power, to be wirelessly transmitted to the ground. In large-scale emergencies, it can be expected that terrestrial sources of electrical energy will also be damaged; thus an intermediate power source is needed, which can be supplied with the help of a high-flying airship.

Navigable airship: In this design, power in the form of laser energy will be sent from SEO solar power satellites to an intermediate platform hovering high in the stratosphere. These dirigible-type airships, powered by solar power, are designed to continuously operate at 60,000–100,000 ft above the disaster area for weeks to months as needed, with the capability to receive laser power energy from space and relay up to 1 gw (one billion watts) of energy to Earth's surface via laser power or cloud-penetrating microwave beams. The 1 gw is sufficient to power a million homes during a crisis, matching the capacity of a coal or nuclear power plant. Portable, expanding receptor antennas can be erected on site to receive this energy with the purpose of running generators or beefing up the existing electrical grid.

Emergency communications: When a devastating hurricane hits, one of the greatest constraints in providing relief will be the lack of communications. Phone towers for mobile telephony will often be knocked out, slowing the local team's ability to coordinate relief efforts. In this design, the same airship providing power will be equipped to serve as a tall multi-purpose telecommunications tower, filling in as a relay and hub for telecommunication services.

Search and rescue sensors: Such airships can also be equipped with passive electro-optical (EO) and active radar sensors allowing rescue managers to quickly scan the debris and locate people trapped in the aftermath of the disaster. This task can be accomplished in a fraction of the time it would take to find them in other ways.

The brief concludes that, in the event of a disaster, solar power satellites have an important role to play in saving lives as well as restoring order.

With access to space-based solar power produced by Sun-synchronous satellite networks, rescue agencies will be able to direct electrical power to any location on the planet. Although still in the planning stages, this technology is paving the way for an alternative power grid that can be used to the benefit of all (Power 2011).

Concluding Thoughts

There is no way to foresee precisely the areas in which Sunsat products/services will be in greatest demand, for some space energy applications will be as broad as charging the batteries of cell phones, laptops and other new media devices while providing roaming connectivity to the Internet in all parts of the globe day or night; or they will be as narrowly targeted as servicing an advance guard of a military operation where access to electrical power is unavailable. All of these are possibilities.

References

Athens Messenger. 2010. Region poised to reap employment from giant solar farm on strip-mined land. http://www.athensmessenger.com/news/local/article_de5725fa-d20f-11df-af4a-001cc4c03286.html. Accessed 7 October 2010.

Davis, D.E. 2011. Hybrid space & near-space solar power disaster relief emergency power, illumination & communication. Proposal submitted to the Ohio University SPS Creative Visualization Project.

Hsu, F. 2010. Harnessing the Sun: Embarking on humanity's next giant leap. *Online Journal of Space Communication*. http://spacejournal.ohio.edu/issue16/hsu.html. Accessed 20 May 2011.

Landis, G. 2004. Reinventing the solar power satellite. National Aeronautics Administration. http://www.nss.org/settlement/ssp/library/index.htm. Accessed 26 May 2011.

Mankins, J. 1997. A fresh look at space solar power: New architectures, concepts and technologies. 38th International Astronautical Federation, NASA, 1997. IAF-97R.2.o3.

Marshall, J. 2009. Space solar power: The next frontier? Pacific Gas & Electric Company. http://www.next100.com. Accessed 13 April 2009.

National Space Society. 2008. Space solar power: An investment for today, an energy solution for tomorrow. *Ad Astra*, *20*(4), p. 50. http://www.nss.org/adastra/volume20/v20n4.html. Accessed 25 May 2011.

Ohio University GRID Lab. 2010. For animation and related content see http://spacejournal.ohio.edu/issue17/present.html.

Potter, D., M. Bayer, D. Davis, A. Born, D. McCormick, L. Dorazio, & P. Patel. Space solar power satellite alternatives and architectures. 2009. AIAA Aerospace Sciences Meeting, Orlando, Florida, 5–8 January 2009.

Power, Illumination, & Communications. 2011. The basis for the Sunsat visualization and technical brief provided by Ohio University students in Issue No.17: SPS Creative Visualization, Online Journal of Space Communication. http://spacejournal.ohio.edu/.

Tobiska, K. 2009. Vision for producing fresh water using space power. American Institute of Aeronautics and Astronautics. AIAA-2009-6817.

Tobiska, K. 2011. Vision for producing fresh water using space power. *Online Journal of Space Communication*. http://spacejournal.ohio.edu/issue16/tobiska.html. Accessed 25 March 2011.

Wang, U. 2009. Solaren to close funding for space solar power. *Green Tech Media*. http://www.greentechmedia.com/. Accessed 1 December 2009.

Woodcock, G. 2010. Solar power satellites: A brief review. International Space Development Conference-Chicago, May 2010. http://sunsat.gridlab.ohio.edu/node/9. Accessed 15 May 2011.

Chapter 3
What Will Sunsats Look Like?

Abstract This chapter suggests several strategic designs for future Sunsats, to include substantially larger photovoltaic arrays in space, solar concentrators, energy converters, wireless power transmitters and power beaming. Technical feasibility and some key technology challenges are addressed, including suitable orbits for Sunsat placement and managing the space environment.

Technical Feasibility

First, let's ask the basic question: is this a workable idea?

Space scientist Feng Hsu has been a proponent of energy development in space for most of his NASA career. Dr. Hsu worked at Brookhaven National Laboratory, where he was a research fellow in such areas as risk assessment, safety and reliability and mission assurances for nuclear power, space launch and energy infrastructure. He is now an even stronger advocate of space power in his role as senior vice president of systems engineering and risk management with the Space Energy Group, a commercial enterprise focusing on renewable energy. This company's website notes that it intends to be "the world's first private enterprise to successfully commercialize space-based solar power (SBSP)" (Space Energy Group 2011).

In 2010, the *Online Journal of Space Communication* asked Dr. Hsu: "Is solar energy from space technologically feasible?" His answer was "positively and absolutely" yes, although he qualifies his reply by explaining that one of the reasons <1% of the world's energy currently comes directly from the Sun is because of high photovoltaic cell costs and PV inefficiencies in converting sunlight into electricity.

Based on existing (terrestrial) technology, a field of solar panels the size of the state of Vermont will be needed to power the electricity needs of the whole United States, and to satisfy world consumption will require some 1% of the land used for agriculture worldwide. Hopefully this will change when breakthroughs are made in conversion efficiencies of PV cells and in the cost of producing them, along with more affordable and higher capacity batteries (Hsu 2010).

Dr. Hsu, who gave permission for his written responses to the Journal to be quoted here, notes that roughly 7–20 times less energy can be harvested per square meter on Earth than in space, depending on location. Likely, this is a principal reason why space solar power has been under consideration for more than 40 years. To be historically correct, as early as 1890 Nikola Tesla, inventor of wireless communication, was writing about and seeking to demonstrate the means for broadcasting electrical power without wires. Tesla later addressed the American Institute of Electrical Engineers regarding his attempts to demonstrate long-distance wireless power transmission over the surface of Earth. He said, "Throughout space there is energy. If static, then our hopes are in vain; if kinetic—and this we know it is for certain—then it is a mere question of time when men will succeed in attaching their machinery to the very wheel work of nature" (Tesla 1892).

Dr. Hsu noted that Dr. Peter Glaser first developed the concept of continuous power generation from space in (Glaser et al. 1968). "His basic idea was that satellites in geosynchronous orbit would be used to collect energy from the Sun. The solar energy would be converted to direct current by solar cells; the direct current would in turn be used to power microwave generators in the gigahertz frequency range. The generators would feed a highly directive satellite-borne antenna, which would beam the energy to Earth. On the ground, a rectifying antenna (rectenna) would convert the microwave energy to direct current, which, after suitable processing, would be fed into the terrestrial power grid."

In describing what a typical solar power satellite (SPS) would look like, Dr. Hsu said the satellite would host a solar panel area of about 10 km^2 in size, and a space-to-Earth transmitting antenna of about 2 km in diameter. On the ground, a rectenna would be constructed about 4 km in diameter corresponding to the expected size and density of the energy beam. Such an installation could yield more than 1 gw of electric power, roughly equivalent to the productive capability of a large-scale nuclear power station.

In summary, he wrote, two critical aspects have motivated research into SPS systems placed in GEO orbit (1) the lack of attenuation of the solar flux by Earth's atmosphere, and (2) the 24-hour availability of space energy (except around midnight during the predictable periods of equinox).

Commercial Viability

Among the key SPS technology techniques are microwave generation and transmission, wave propagation, antennas and measurement calibration and wave control. Dr. Hsu calls these "radio science issues" that cover a broad range of topics, including the technical aspects of microwave power generation and transmission, the effects on humans and potential interference with communications, remote sensing and radio-astronomy observations.

"Is SPS a viable option? Yes, in my opinion, it can and should be a major source of baseload electricity generation powering the needs of our future," states Dr. Hsu, satisfied that SPS meets each of the key criteria except for cost, which is increased by current space launch and propulsion technology. He continues:

> We all know that the expense of lifting and maneuvering material into space orbit is a major issue for future energy production in space. The development of autonomous robotic technology for on-orbit assembly of large solar PV (or solar thermal) structures along with the needed system safety and reliability assurance for excessively large and complex orbital structures are also challenges. Nevertheless, no breakthrough technologies or any theoretical obstacles need to be overcome for a solar power satellite demonstration project to be carried out.
>
> Our society has repeatedly overlooked (or dismissed) the potential of space-based solar power. The U.S. government funded an SPS study totaling about $20 million in the late 1970s at the height of the early oil crisis, and then practically abandoned this project with nearly zero dollars spent up to the present day. A government-funded SPS demonstration project is overdue.
>
> What I really want to point out here is that we can solve the cost issue and make solar power satellites a commercially viable energy option. We can do this through human creativity and innovation on both technological and economic fronts. Yes, current launch costs are critical constraints. However, in addition to continuing our quest for low cost RLV (reusable launch vehicle) technologies, there are business models for overcoming these issues.

Dr. Hsu notes that several such models have been studied and are now being pursued by such aerospace entrepreneurial companies as the SE (Space Energy Group) and the SIG (Space Island Group) based in Switzerland and California. The SE approach is based on systematic development of solar technologies for terrestrial and for space environment applications. The company expects to rely on extensive terrestrial solar technology development as the stepping stone, focusing on the space-grade thin film PV technology innovations for launch cost reductions. The SIG idea is to use and/or modify legacy components of the space shuttle, turning the huge volumes of the external shuttle tanks into a commercial asset for the space-based research and orbital tourism industry. Increased demand in space tourism will certainly bring about a greater number of launches, which should drive down space transportation costs.

From Feng Hsu's perspective, solar power satellites are technically feasible—and may be economically achievable sooner than was thought.

New Architectures

From 1995 to 1997, NASA conducted a re-examination of the technologies, systems concepts and terrestrial markets that might be involved in future space solar power systems. This study was reported by NASA scientist John C. Mankins, who worked for the agency's Advance Projects Office. A summary of the study, its goals and findings, is instructive for our consideration of Sunsat feasibility.

According to Mankins (1997), the principal objective of this "Fresh Look" study was to determine whether a solar power satellite (SPS) and associated systems could be defined that could deliver energy into terrestrial electrical power grids at prices equal to or below ground alternatives in a variety of markets, that could do so without

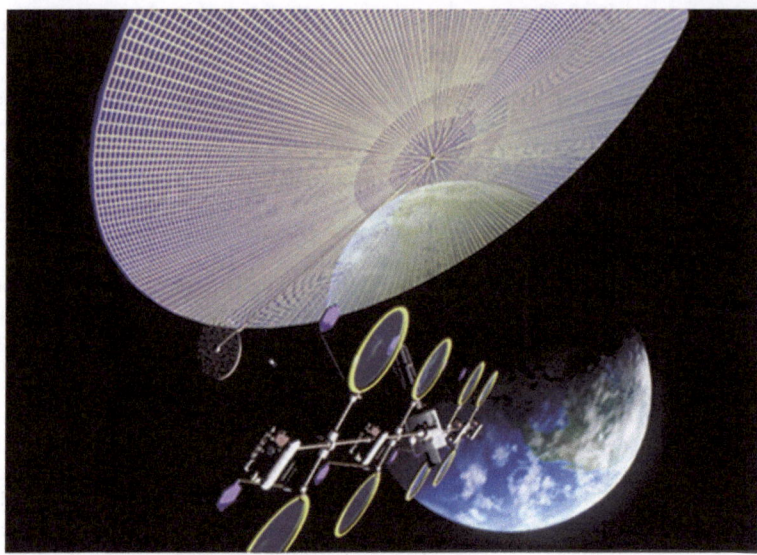

Fig. 3.1 Pictured is a NASA image of a solar power satellite called the "solardisk," expected to generate as much as 5 gw of electricity in space (NASA 1999)

major environmental drawbacks, and that could be developed at a fraction of the initial investment projected for the SPS Reference System of the late 1970s.

Approximately 100 experts in a wide variety of disciplines participated in this 2-year study, which involved three major workshops. Working within the global energy marketplace of the twenty-first century—including a major focus on emerging nations—the study examined five different markets and about 30 different SPS concepts, ranging from the 1979 SPS Reference Concept defined by the U.S. Department of Energy and NASA to very advanced concepts involving technologies that have not yet been validated in the laboratory (Fig. 3.1).

Following a preliminary assessment of technical and economic risks and projected costs, Mankins reported, "Seven SSP system architectures and four specific SPS concepts were chosen for examination in greater depth using a comprehensive, end-to-end systems analysis employing a desktop computer modeling tool developed for the study…. Several innovative concepts were defined and a variety of new technology applications considered, including solid state microwave transmitters, extremely large tension-stabilized structures (both tethers and inflatable structures), and autonomously self-assembling systems using advanced in-space computing systems."

A key strategy to achieve initial cost goals was to avoid wherever possible the design, development, test and evaluation costs associated with SSP-unique infrastructure, such as fully reusable, heavy-lift launch vehicles. Three architectures in particular were identified as promising: a Sun-synchronous low Earth orbit (LEO) constellation, a middle Earth orbit (MEO) multiple-inclination constellation, and one or more stand-alone geostationary Earth orbit (GEO) SPS serving single, dedicated ground sites.

The study gave attention to what had changed to make it possible in 1997 and thereafter to consider implementing space-based systems for energy production. Mankins wrote that the most important contextual change was the increasing demand for energy globally and the growing concern regarding carbon combustion, CO_2 emissions and global climate change. As a result, a major priority was being placed on the development of renewable energy sources.

He also noted a change in U.S. national policy that called "for NASA to make significant investments in technology (not a particular vehicle) to drive the costs of ETO (Earth to orbit) transportation down dramatically. This is, of course, an absolute requirement of space solar power." Such a policy, he said, was independent of any SSP (space solar power)-related considerations and thus need not be "charged" against the cost of developing SSP technology.

The Mankins study concluded that there had emerged "a new paradigm for the relationship between governments and industries, for example with NASA's role in research and development to reduce risk and to seek government mission applications—but not to actually develop operational systems" (Mankins 1997, p. 8).

Newer Research

In 2004, Geoffrey Landis of the NASA Glenn Research Center in Berea, Ohio, published findings entitled "Reinventing the Solar Power Satellite." The NASA Glenn team was looking at new designs for a space solar power system that would provide electrical power at more economically competitive rates. As the perceived cost of space solutions was thought to be a barrier, their approach was to examine (and create the conceptual designs for) more practical approaches to space power production and delivery (Landis 2004, p. 1).

Three new concepts for solar power satellites were invented and analyzed. The concepts included (1) a solar power satellite positioned in a higher orbit (e.g., the Earth-Sun L2 point), (2) a solar power satellite in GEO orbit with no moving parts, and (3) a GEO non-tracking solar power satellite with integral-phased array. The integral-array satellite had several advantages, including an initial investment cost approximately eight times lower than the conventional design. The related details of these approaches, including their disadvantages, can be found in the paper.

The Landis paper concluded with the following observations: "A space solar power generation system can be designed to work in synergy with ground solar power. Previous space solar power architectures were designed to deliver 24-h power; this design constraint was relaxed. A non-tracking, integrated solar/microwave space power system can be configured to match peak power demand." The team reported that minimum system sizes decrease in power by a factor of 8 with face-on solar array, and by a factor of 4 with a 4 PM/8 AM tilt, and observed that the ground rectenna scaled proportionately (Landis 2004, p. 30).

Several findings of this study were thought to be helpful in accelerating Sunsat implementation. One of the most noteworthy is the integrated solar/microwave design, which makes this approach considerably more feasible (than tracking system concepts) by decreasing the investment required to reach first power.

Other Technical Challenges

Belvin, Dorsey and Watson, researchers at NASA Langley Research Center in Hampton, Va., drew important lessons from their studies on "very large in-space structural systems" (Belvin et al. 2010). In the 1990s, this team worked to address some of the problems of introducing and managing mechanical systems in space. Seeking innovative solutions that might lead to more economically and technically feasible designs for solar power satellites, the researchers tackled four big issues: modularity, material systems, structural concepts and in-space operations. They proposed a "building-block" approach in two phases.

The first phase focuses on a near-term application/customer of SSP that is willing to pay a premium for consistent and uninterrupted power. An example of this would be military bases in remote and hostile regions, where the logistics train for fuel (to run generators) is very expensive, dangerous and subject to constant disruption. Low power SSP systems may also be used in orbit around the Moon, Mars and other Solar System planets and moons to provide power to surface rovers and outposts. The power generation level (at the source) for this first-phase application might be from 100 to 5,000 kw. This application would use current and pending technology (structures, solar cells, ion propulsion/station keeping, avionics and power beaming) for spacecraft subsystems and automated rendezvous and docking for spacecraft assembly.

The goal of the second phase would be to develop the advanced technologies required for SSP systems capable of producing 100–2,000 mw of power for commercial transfer to Earth's power grid. Such large satellites would be developed only when appropriate systems and technologies were sufficiently advanced to make them commercially viable. Using block upgrades on first-phase systems to develop and demonstrate the advanced technologies as they become available would reduce the cost, schedule and performance risks of very large system implementation. In addition, the probability of commercial system development success would be maximized because development would not begin prematurely (Belvin et al. 2010, p. 3).

The team noted that the successful first phase SPS (solar power satellite) demonstrator would serve to validate performance/economics models and operations experience that would permit large-scale system architectures to be developed for a 1 gw-class SPS. The choice of wireless power transfer technology, specifically the wavelength (RF or laser), would influence the SPS antenna size and thermal requirements. Large inflatable concentrators have been proposed as a way to reduce the photovoltaic area (and its cost), they said, but little attention has been paid to long-term space durability.

The specific advances in MSMS (materials, structures and mechanical systems) technology thought to benefit very large space structural systems in the coming decades include:

- Modularity (module based assembly and upgrade)
- Material systems (space durable, high temperature, and thin films)
- Structural concepts (inflatable, rigidizable and gossamer concepts)
- In-space operations (deployment, assembly and repair)

In terms of in-space operations, the team suggested the need for multiple launches to place the subsystems into LEO, where the system could be assembled and then transferred to its final orbit, or all of the subsystems could be transferred to final orbit and final assembly performed there. They wrote, "Either approach will require a robust set of in-space operational capabilities, including automated rendezvous, docking and berthing, assembly, and servicing and repair. Recent robotics missions have significantly matured, the key in-space operations technologies needed for SSP" (Belvin et al. 2010, p. 6). They concluded that technology advances in all four areas over the last 15 years make the technical feasibility of an operational SPS system much greater than it was just two decades ago.

Corporate Research

The Boeing Company has maintained an interest in SSP for more than 40 years. Several space scientists and engineers who were (or are still) employees of Boeing have spent almost their entire careers working on solar power satellite concepts, technologies and applications. In 2008, the Boeing team working in El Segundo and Huntington Beach, California, published an overview of space solar power satellite alternatives and architectures. Among the summary conclusions reported were:

> NASA and industry have studied (SSP) intermittently during the 1990s and early 2000s. System sizes are huge (solar arrays several thousand meters across; power levels of thousands of megawatts). Due to the divergence of the microwave beam, a large amount of power must be collected to achieve an economically recoverable power density at the receiver array.
>
> Notional SSP cost modeling assumed that the technology would be mature to 2028, that each satellite would be capable of a 30 year mean mission duration, that a 97% learning curve would be needed for production, that (if launched from moon or Earth) infrastructure, materials and factory for production would already be in place, and that robonauts (assembly robots) would be included in production costs for maintenance to be launched with the satellite.
>
> Projections for global electric power generation capacity suggest that "to meet 10% of future global electricity requirements, 20 GW of SSP capacity must be deployed per year" (Potter et al 2009).

Concluding Thoughts

The new solar power satellite industry will position above Earth new types of energy infrastructure hosting many of the features of communications platforms, including a satellite bus (physical structure), solar arrays, onboard processing, telemetry control and wireless transmission systems. Unlike the comsats that gather only a modest amount of the Sun's radiation to power their spacecraft, the Sunsat antennas will be collecting and concentrating energy for the principal purpose of relaying it to Earth as its predominant service and product.

Whenever technological developments lead to thinner, lighter, cheaper photovoltaic (PV) cells that make terrestrial power production more efficient, those same benchmarks also benefit comsat systems in space. For solar power satellites, these same advancements will have a multiplying benefit many times greater. This difference alone may make the launch of Sunsats possible sooner than later, since space producers of energy are looking to reduce the mass and increase the productivity of their antennas. Antenna size and weight are key to holding down costs of launching their considerably larger collector arrays into space.

Also benefitting each of these space industries will be promising new developments in remote construction, assembly, repair and replacement. Among the more innovative Sunsat-related designs are architectures that consist of more than one satellite, networking them together within a common space orbit, creating a larger photovoltaic mass. Such satellite systems may one day be interlinked for global service.

References

Belvin, W. K., J. T. Dorsey, & J. J. Watson. 2010. Solar power satellite development: Advances in modularity and mechanical systems. *Online Journal of Space Communication*. http://space-journal.ohio.edu/issue16/belvin.html. Accessed 1 May 2011. Adapted from a paper delivered to the International Symposium on Solar Energy from Space, Toronto, Canada, September 8–10, 2009.

Glaser, P. G., F. P. Davidson, & K. I. Csigi. 1968. *Solar power satellites—The emerging energy option*. New York: Ellis Horwood.

Hsu, F. 2010. Harnessing the sun: Embarking on humanity's next giant leap. *Online Journal of Space Communication*. http://spacejournal.ohio.edu/issue16/hsu.html. Accessed 22 February 2011.

Landis, G. A. 2004. Reinventing the solar power satellite. NASA Glenn Research Center, NASA/ TM-2004-212743. http://gltrs.grc.nasa.gov/cgi-bin/GLTRS/browse.pl?2004/TM-2004-212743. html. Accessed 15 May 2011.

Mankins, J. C. 1997. A fresh look at space solar power: New architectures, concepts and technologies. Lecture given at the 38th International Astronautical Federation. 1AF-97-R.2.03.

NASA. 1999. (reported by) http://www.environmentalgraffiti.com/featured/solar-energy-beamed-down-space-lasers/17556. Accessed September 12, 2011.

Potter, D., M. Bayer, D. Davis, A. Born, D. McCormick, L. Dorazio, & P. Patel. Space solar power satellite alternatives and architectures. 2009. AIAA Aerospace Sciences Meeting, Orlando, Florida, 5–8 January 2009.

Space Energy Group. 2011. Corporate Overview. http://www.spaceenergy.com/About/About.htm. Accessed 25 May 2011.

Tesla, N. 1892. Experiments with alternate currents of high potential and high frequency. Address to the Institution of Electrical Engineers, February 1892, London, England. http://www.tfcbooks.com/tesla/1892-02-03.htm. Accessed 16 May 2011.

References

Chapter 4
How Will Sunsats Be Delivered to Space?

Abstract This chapter outlines several approaches to delivering powersats into low, medium, geosynchronous, Sun-synchronous and other space orbits. A historical context is given and next-generation launch strategies are introduced. Increased spacecraft size, mass and deployment frequency of payloads and deployment are among the challenges discussed.

Launching Sunsats

As with communications satellites, solar power satellites must be lifted from Earth and delivered into designated orbits. Some will be positioned quite near Earth, while others will be farther away. To place any satellite in space for the purpose of relaying energy to the ground, providers of these services must go through a prior approval process with the International Telecommunications Union and other oversight authorities.

The more promising locations for directing power to Earth appear to be in LEO at roughly 300 km, in the geosynchronous Earth orbit (GEO) at 36,000 km or in an elliptical orbit that will permit always-in-the-Sun reception. Some strategists propose using space-to-space energy reflectors to relay power from satellites gathering the Sun's rays in daylight, transferring power to satellites orbiting in the shadow of Earth from where the beam will be down-linked to ground antennas.

Others look to the Moon as a future base for collecting and beaming solar power to Earth. Such an energy source could be used as well for the electric propulsion of spacecraft into deeper space. Among the more innovative Sunsat architectures are those that network multiple solar power satellites, treating them as a single photovoltaic mass serving one or more than one world region.

An Historical Perspective

Space engineer Ralph Nansen has spent much of his career designing, developing and advocating concepts that relate to space solar power. Starting as a designer on the Bomarc rocket-powered missile for the Boeing Company, Nansen was selected in 1961 to design the configuration used by Boeing in building the giant first stage of the Saturn V Moon rocket. In 1962, he became design manager of the Saturn S-1C fuel tanks, the first stage of the rocket that propelled the *Apollo* astronauts to the Moon.

From 1975 to 1980, Nansen served as Boeing solar power satellite program manager. He gathered the team of engineers, scientists and associate contractors that developed the overall SPS concept under the auspices of the Department of Energy and NASA. He presented numerous papers and participated in international conferences on future space projects in Germany and Egypt. He was invited to China as a member of the first Space Technology Exchange Mission in 1979. Nansen was asked to testify before such Congressional committees as the Senate Space Subcommittee in 1976 and the House Subcommittee on Space and Aeronautics in 1978 and again in September 2000.

From 1985 to 1987, he was responsible for developing the design proposal for a fully reusable horizontal take-off space transportation system and the structural design of Boeing's National Aerospace Plane concept. Nansen retired from Boeing in 1987 and has since written two books on the world energy crisis and potential solutions from space.

Nansen says the barrier to SPS development is the lack of a low-cost space transportation system for launching the satellite hardware. "Without a reusable launch system there is little hope of deploying a significant capability to generate competitive cost electric energy from space. The problem is not technology; it is the up-front investment money and understanding of what is required" (Nansen 2010).

In his article for the *Online Journal of Space Communication* on the topic of low cost access to space, Nansen focuses on the specifics of developing a space transportation system based on reusable vehicles, an approach that he is confident will finally make solar power satellite deployment commercially viable. The first step, he writes, "is to look at what has occurred in the past and see what has happened, and why it happened. To make the right choices for the future… we need to understand what is different now." He continues:

> All of the early launch systems starting with the launch vehicle for *Sputnik* were expendable rockets. In the early days, there wasn't much choice. To reach orbit, launch systems had to be made as light as possible to achieve orbital velocity. There was nothing left over for adding recovery systems that would allow reuse. As time went on, systems got more efficient, but overall program cost became a key decision maker. To minimize cost, payload was reduced. The added cost of development for a reusable system was traded against the number of flights required. The other element was that many of the payloads needed to go to high orbits that made the recovery of the upper stages difficult and costly. As a result, the market was not large enough to justify the cost of a reusable system. The optimum manageable design was always to build a highly efficient expendable system. Once the commercial satellite providers managed to become profitable using expendable rockets, the launch vehicle builders had no real incentive to develop reusable systems (Nansen 2010).

"As the Saturn/Apollo Program was winding down," Nansen writes, "NASA stepped forward with a bold plan that could have led to a new era of space development. It was the plan for a space shuttle. NASA's criterion was for a fully reusable two-stage winged vehicle that would burn liquid hydrogen and oxygen as the propellants in both stages." The big constraint, he says, was the level of technology available in 1970. The two biggest stumbling blocks were (1) the maximum gross liftoff weight and (2) the need to use hydrogen as the booster fuel. Hydrogen fuel use dictated a much larger vehicle than would be required with a hydrocarbon fuel booster. The gross lift-off criterion was incompatible with hydrocarbon fuel and the size of a hydrogen fueled booster. None of the bidding contractors could meet the liftoff criteria.

"Now close to 40 years later," he writes, "the United States has had two fatal accidents on space shuttle flights, each mission costs a small fortune to fly, and now the entire fleet is slated to be retired.... The question is: What can we do today to develop a reusable space transportation system with a minimum of developmental costs?"

Nansen's recommendation is "to reach back 40 years to the technology we understand, update it with modern knowledge and materials and incorporate what is learned into a fully reusable vehicle that applies the known principles of low cost transportation systems. Those principles are high usage, low maintenance, reasonably sized payloads, and ease of loading and unloading. When a transportation system reaches maturity with these characteristics, the cost of operating the system can be expected to be between three and five times the cost of fuel. With today's systems, the cost is over a thousand times" (Nansen 2010).

With the development of a fully reusable launch vehicle designed for commercial use by people who understand commercial operations, Nansen believes that solar power satellite hardware can be launched at a low enough cost that the satellites will provide competitively priced electricity to Earth. "Such an event would be the beginning of the new era of energy from space that would bring economic growth to the world while at the same time stopping the addition of carbon dioxide to our atmosphere" (Nansen 2010).

Launch Strategies

It can be assumed that any solar power satellites built today will be launched on the same private, commercial and government rockets used by the comsat industry to lift their communications satellites. It can also be assumed that, as cheaper and more suitable launch options appear, both Sunsat and comsat clients will benefit.

Forty or more years of practice has led to a high level of confidence in the launch industry's capability to deliver spacecraft into orbits of choice, using a range of launch vehicles to accommodate quite specialized payloads. The prospect of a new generation of satellites pursuing a new business category—that is, providing a continuous supply of clean and abundant energy to all countries—will give the launch industry the spurt of growth it has been hoping to see. Launching solar power satellites will be its first

Fig. 4.1 The Falcon heavy
launch vehicle of Space
Exploration Technologies
is to be launched at Cape
Canaveral in 2014. The
rocket will lift satellites
and cargo weighing 53 t
into low Earth orbit at
200 km (SpaceX 2011)

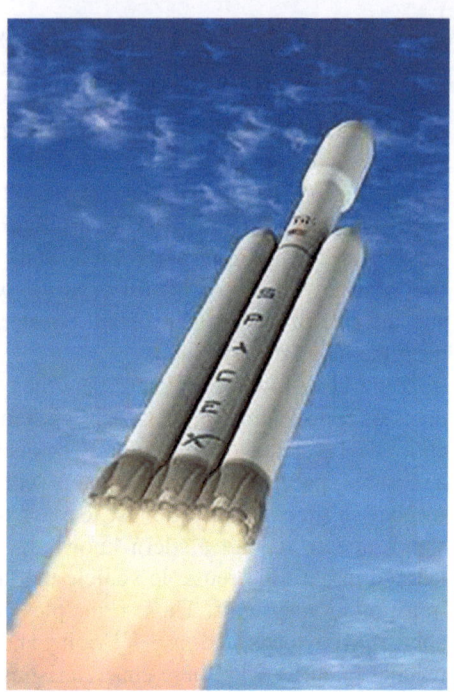

opportunity to demonstrate that it can provide not only safe and reliable transport to space, but also can deliver it in sufficient volume and at sufficiently low cost to ensure the worldwide availability of competitively priced electricity (Fig. 4.1).

Bruce Elbert, in his widely used *Introduction to Satellite Communication*, points out that the three most common criteria in launch vehicle selection relate to launch mass capability, the reliability or success record of the system and the cost of use (Elbert 1999, p. 406). Spacecraft are normally designed for compatibility with a particular launch vehicle to be placed into a specified orbit. The place where a spacecraft is launched, whether on land, sea or in the air, will very much depend on its ultimate destination. For example, a GEO placement in space will prompt a launch location closest to the equator, since the highly desired GEO orbit is 36,000 km above Earth's equator. For a spacecraft with a non-GEO destination, launch will likely occur from a site located at some higher latitude.

"The sequence of steps that begin when a spacecraft aboard its launch vehicle leaves the launch platform and concludes when the spacecraft is separated in space is called the launch mission. In some cases the launch mission is completed short of the actual orbital destination when, for reasons of cost or complexity, the spacecraft is unloaded and caused to continue to the designated altitude and position using its own power. This is most often the case with GEO satellites, when the launch vehicle places its payload into a geo-transfer orbit (GTO). In other cases, the launch vehicle accompanies the payload the entire distance" (Elbert 1999, p. 406).

Fig. 4.2 China's powerful
Long March-5 rocket in
development will sport
engines with the thrust
of 120 t, with a test launch
scheduled for 2014
(Zak 2010)

Some plans involve assembling solar satellites and their antennas from components lifted by medium power rockets into LEO, possibly using the International Space Station or other space platform as a staging area, later transferring them into their final position in a geosynchronous or other orbit. Other plans include inserting the solar spacecraft and its large arrays directly into orbit using more powerful and agile thrusters (Fig. 4.2).

Reducing Costs

Phillip Chapman, an Australian-born geophysicist and astronautical engineer who served as a scientist-astronaut for NASA during the Apollo era, wrote about economical launch vehicles, energy and environmental policy and space solar power in Issue No. 16 of the *Online Journal of Space Communication*. Giving thought to the cost of launching solar power satellites and incorporating launch technologies available today, he concluded that the cost of spaceflight is not a serious impediment to realizing the advantages of power from space.

"It is important to recognize that spaceflight is not intrinsically expensive," Chapman notes. "The energy needed to place a payload in LEO is ~12 kWh/kg.

If it were possible to buy this energy in the form of electricity at U.S. residential prices, the cost would be <$1.30/kg. Rockets are very inefficient, but the cost of the propellants needed to reach orbit is typically <$25/kg of payload.

"The principal reason that launch to LEO is currently so expensive (>$10,000/kg) is that launches are infrequent—and they are infrequent because they are so expensive. Launch vehicles (LVs) are costly to build because the production volume is low; each LV is thrown away after one use. Annualized range costs are shared among just a few launches, and the staff members needed for LV construction and launch operations are grossly underemployed. The quoted prices for launch would be much higher still were it not that in most cases the Department of Defense or NASA has absorbed the LV development cost" (Chapman 2010).

He calculates that economies of scale in any significant space-based solar power (SBSP) program will permit launch at acceptable cost, even without major advances in launch technology. "To be definite, a fairly modest SunSat deployment program is assumed, with the first launch taking place in 2015, leading to an installed SunSat capacity of 800 gwe in 2050. This goal will represent somewhere between 6% and 9% of the total global capacity that we will need by then" (Chapman 2010).

Chapman's analysis uses simple standard models to approximate the performance and cost of LVs, with subsystem characteristics comparable to those of existing engines and vehicles. "The only major technical innovation considered," he writes, "is the introduction of reusable LV stages, and the only major change in spaceflight practice is launch from an equatorial site." There was no attempt, he states, to optimize the launch architecture, although improved designs and advanced technologies would offer significantly lower costs (Chapman 2010).

The principal problems in closing the business case for a launch services provider that supports space-based solar power, he says, are related to financing the venture rather than the cost of operations or the eventual profitability. For example, he notes: "[A] launch price of $450/kg leads to a maximum deficit of $60 billion in the 12th year of the deployment schedule, and the cumulative cash flow does not become positive until the 22nd year—but the end result in 2050 is a profit of $180 billion (Chapman 2010).

"The delay in profitability exceeds the planning horizon of most venture capitalists, so the project probably requires both a strong government commitment to completing the deployment as well as some form of financial guarantee. Creative financing could help; for example, the launch price could be set at $600/kg in the early years, with a contractual obligation to refund some of the money once the cash flow went positive" (Chapman 2010).

Chapman isn't recommending a particular design for RLVs; rather, in this paper, his purpose was "to show by example that the cost of launch to LEO is not a reason to delay implementation of SBSP as a major contributor to energy supply in the United States and around the world. The need is urgent, and the best time to begin a serious development program is right now" (Chapman 2010).

Gordon Woodcock, honored in 2011 by the National Space Society for distinguished service in advancing the case for space-based solar power, has addressed the topic of launch costs on multiple occasions. He calls launch costs "The Big Show-Stopper" (Woodcock 2010, p. 1).

In a presentation at the 2010 International Space Development Conference in Chicago, Woodcock concluded that re-usable systems can deliver acceptable costs if (1) there is high demand; (2) these systems have long life; (3) there is a short turnaround time; and (4) they have modest turnaround cost. His analysis shows fully reusable vehicles are not worth the investment unless demand is at least 50–100 launches per year, and that the turnaround is less than a week on the ground between flights.

For getting started, he said, investment analysis shows a partially reusable heavy lift vehicle with flyback booster can be justified at 3–5 launches per year or more (when there are additional purposes for such missions as human space exploration). He assumes that the smaller, fully reusable passenger vehicles for space tourism to orbit are helpful steps along the way (Woodcock 2010, p. 3).

Reusable Rockets

The National Space Society gave its Space Pioneer Award for Business Entrepreneur to SpaceX in 2011, in recognition of its successful launch of two Falcon 9 rockets and the safe return of its *Dragon* capsule. NSS Executive Committee Chair Mark Hopkins noted, "The high cost of launch has always hampered the exploration and development of space. With its Falcon Heavy vehicle, SpaceX seeks to achieve a major reduction in launch costs. Such a reduction could enable entirely new categories of space industry" (Hopkins 2011).

SpaceX CEO Elon Musk announced in April 2011 that the company had scheduled two or three Falcon 9 launches for 2011, with launch rates ramping up to five or six in 2012, growing to 12 per year by 2014. Musk said the company's goal is to launch this vehicle 20 times per year, a rate that would permit SpaceX to further reduce per-launch charges (de Selding 2011).

Musk said the company's Falcon 9 rockets would be entering into competition with the Atlas 5 and Delta 4s for U.S. Air Force contracts, but would also compete with the Russian Proton and the European Arianes in the commercial marketplace. When measured in terms of the cost of placing a given satellite into orbit, he said, the Falcon 9 Heavy would be only half as expensive as the Russian Proton (de Selding 2011).

NASA spokesperson Lori Garver was quoted in a *Space News* article as saying that a conventional NASA procurement of its own heavy-lift rocket, including its first flight, would cost nearly $4.5 billion. Outsourcing development to SpaceX, she said, would cut that figure by 60%, but only if other customers purchased the vehicle, permitting scale economies to reach maximum effect (de Selding 2011).

China's launch industry will feel the impact of these developments as well. According to *Aviation Week & Space Technology* editor Frank Morring, "Executives at China Great Wall Industry Corp. find it hard to believe that U.S. Space Exploration Technologies, Inc. (SpaceX) is offering lower launch prices than they can. But they concede privately that it's true."

Morring goes on to explain, "China Great Wall, the marketing arm of China Aerospace Science and Technology Corp. (CAST), is opening a one-person office in Washington, DC this summer to push Chinese space products, including solar arrays. Chinese officials say they find the published prices on the SpaceX website very low for the services offered, and conceded they couldn't match them with the Long March series of vehicles even if the U.S. export-control regulations made it possible for them to loft satellites with U.S. components in them." The SpaceX website has an advertised lift capacity of 10,450 kg for the Falcon 9 payloads from Cape Canaveral for $54 to $59.5 million (Morring 2011, p. 22).

Alternative Approaches

Multiple strategies abound for lifting people and material into space more efficiently, more often and less expensively. One of the less talked about strategies is to use highly focused laser or microwave power to lift satellite vehicles, their parts or payloads into LEO; another is the related space elevator. A common version of the space elevator involves connecting a high strength ribbon (a carbon nanotube tether) from a space satellite to an offshore sea platform. Mechanical lifters attached to the ribbon would be propelled up the ribbon, pushing cargo into space.

Dallas Bienhoff, in a 2008 paper presented to the AIAA, touched on some of these alternative approaches. He wrote:

> From the brute force approach to a more elegant, precisely choreographed and integrated system, the Tether Launch Assist approach can place payloads onto a geosynchronous transfer orbit (GTO) trajectory using a smaller launch vehicle and less than half the upper stage propellant compared to our current rocket/upper stage approach. Development costs for the suborbital RLV are reduced relative to typical RLVs due to the lower delta v requirements for launch and the need for smaller upper stages that perform orbit circularization only. Upper stage capability requirement is reduced as the perigee burn function is provided by the tether. Operationally, the launch vehicle carries the payload to altitude and releases it in time to meet the passing tether payload hook. The tether rotates so the capture hook is traveling in the opposite direction as its center of mass when the payload is snatched to minimize the relative velocity between the RLV and capture hook. After snatching the payload from free space, the tether rotation carries it upward to its release position 180° away. Tether design is such that the release velocity equals the perigee velocity required for the payload to reach its desired apogee. An apogee burn is necessary for final orbit circularization.
>
> Space elevators...may offer the ultimate low-cost access to space. Consisting of an Earth station, a ribbon, a climber and a counterweight beyond GEO, space elevators may be able to place payloads into GEO for about $100/kg. The climber has wheels, or grippers, that squeeze the ribbon and drive the carrier up to GEO. The ribbon extends from a counterweight beyond GEO to an operating platform on the ocean's surface along the equator. Lasers beam energy to photovoltaic cells on the climber, which provides the electricity to power the grippers. Depending on climber speeds, trip time to GEO may take anywhere from 1 to 10 days. [Subsequent climbers] can initiate their ascent as soon as the previous one reaches the altitude where gravity has decreased to 0.1 g. Because space elevator

ribbons are one-way paths, each elevator site will need two or more ribbons for efficient operations; one for Earth-bound climbers and one or more for space-bound climbers (Bienhoff 2008, p. 8).

A new and plausibly workable approach to Earth-to-space propulsion calls for heating a rocket's propellant by focusing energy on it from ground-based lasers or microwave sources. This concept to "transmit the energy from the ground to the vehicle" was developed in 1991 by Jordin Kare of Kare Technical Consulting. Instead of explosive chemical reactions onboard a rocket, beamed thermal propulsion would launch a rocket by shining laser light or microwaves at it from the ground (Patel 2011, p. 1).

Beamed thermal propulsion systems would involve focusing the beams on a heat exchanger aboard the rocket. The heat exchanger would transfer the radiation's energy to a liquid propellant such as hydrogen, converting it into a hot gas that is pushed out of the nozzle. Proponents suggest that this approach would make possible a reusable single-stage rocket that has 2–5 times more payload space than conventional rockets, dramatically slashing the cost of sending payloads into a low Earth orbit. NASA is now conducting such a study to examine the possibility of using beamed energy propulsion for future space launches.

Kare had calculated that it would take 8–10 min for a laser to put a craft into orbit, while microwaves would do the job in 3–4 min. The vehicle would have to be designed without shiny surfaces that could reflect dangerous beams, and aircraft and satellites would have to be kept out of the beam's path. Such launch systems would be built in high-altitude desert areas, so danger to wildlife would be minimized (Patel 2011, p. 2).

Concluding Thoughts

Launching satellites safely and economically into space is one of many significant challenges facing the satellite industry. Any positive momentum toward cheaper launches will be good news for space energy, space communications and related space businesses. Private/public initiatives to regularize space transport are helping to establish access to space as a viable enterprise in the way that terrestrial aerospace is viable today.

To avoid the high costs of launching people and cargo into space, some visionaries see space-based infrastructures being built from materials found in space, with robotic manufacturing and assembly managed from Earth via virtual communications and control. Although this seems far off, solar power plants operating in near-Earth orbits can be expected to provide a near-term market large enough to stimulate a more diverse space transportation system. These developments mesh well together. With lower cost space transportation, energy from space becomes the go-to source for supplemental (and eventually replacement) power, the volume of which will drive down overall costs.

References

Bienhoff, D. 2008. Space infrastructure options for space based solar power. 6th International Energy Conversion Engineering Conference (IECEC). 28–30 July 2008. Cleveland, Ohio.

Chapman, P. K. 2010. Deploying SunSats. *Online Journal of Space Communication*. http://space-journal.ohio.edu/issue16/main.html. Accessed 2 May 2011.

de Selding, P. B. 2011. After servicing the Space Station, SpaceX's priority is taking on EELV. *Space News*. http://www.spacenews.com/venture_space/110412-eelv-spacex-priority-after-iss.html. Accessed 28 May 2011.

Elbert, B. R. 1999. Introduction to Satellite Communication. USA: Artech House.

Hopkins, M. 2011. The ISDC Program. International Space Development Conference. Huntsville, Alabama. May 2011.

Morring, Jr. F. 2011. Cut rate. *Aviation Week & Space Technology*. http://bitshare.com/files/gp6o2eb8/Aviation-Week-Space-Technology-April-25.May-2---2011.pdf.html. Accessed 24 May 2011.

Nansen, R. H. 2010. Low cost access to space is key to solar power satellite deployment. *Online Journal of Space Communication*. http://spacejournal.ohio.edu/issue16/main.html. Accessed 2 May 2011.

Patel, P. 2011. Beaming rockets into space. www.astrobio.net.

SpaceX. 2011. 5 April 2011 press release. http://www.spacex.com/.

Woodcock, G. 2010. Solar power satellites: A brief review. Based on a May 2010 International Space Development Conference presentation, Chicago, Ill. Development Conference-Chicago, May 2010. http://Sunsat.gridlab.ohio.edu/node/9. Accessed 15 May 2011.

Zak, A. 2010. China considers big rocket power. BBC News, 26 July 2010. http://www.bbc.co.uk/news/.

Chapter 5
How Will Sunsat Power Be Captured on Earth?

Abstract This chapter addresses both potential opportunities and expressed concerns relating to wireless transmission of space solar energy to Earth in baseload and related electrical power applications. Safety protections associated with the design, location and redistribution of energy on the ground are also discussed.

Future Prospects

A small group of students and faculty at the Georgia Institute of Technology is framing a "space power grid" architecture that will position solar installations in orbit around the globe as a means of exchanging power between terrestrial power plants located in different parts of the world. Building on the revenue from this market, the intent is to then proceed in the construction of the large space power stations that can generate solar electric power for all nations.

This approach, they say, will set up an evolutionary, low-risk, revenue-generating path to "realize the global dream of space solar power." The strategy is incremental, concentrating on helping terrestrial power plants become viable, and then working to align public policy priorities with the goal of a sustainable supply of energy from space.

The team's focus is to help establish a working relationship between the space and energy industries of India and the United States. "Much of humanity today does not enjoy the $0.10/kWhe, uninterrupted delivery of electric power that is taken for granted in urban industrialized societies," the group from Georgia Tech wrote in a paper they presented to the Solar Power Symposium of the 2011 International Space Development Conference held in Huntsville, Alabama (Dessanti et al. 2011, p. 1).

"In regions that are not wired for power, residents pay exorbitant costs for a few watts or watt-hours and suffer lack of basic amenities and opportunities. Thus the first point to make is that competing with the efficient, reliable terrestrial utility and power grid is not the principal purpose of a space-based electric power resource."

D.M. Flournoy, *Solar Power Satellites*, SpringerBriefs in Space Development, DOI 10.1007/978-1-4614-2000-2_5, © Don M. Flournoy 2012

"The ability to reach all parts of the world at any time is a very significant characteristic, beyond being worth a high price. On the other hand, it is entirely possible that the price commanded by terrestrial utilities will keep rising beyond the level where we can make SSP viable even in this market" (Dessanti et al. 2011, p. 1).

Narayanan Komerath, professor at the Daniel Guggenheim School of Aerospace Engineering at Georgia Tech, has been engaged with the idea of a global space power grid since 2006. In a personal communication he wrote, "The notion of exchanging terrestrial power through space is still a complex one for most people to digest, but that is because they are not used to the idea that collaborating with the terrestrial energy community is smarter than trying to articulate why the government should choose one over the other." He thinks a U.S.-India space-based power exchange demonstration would constitute a rational first step toward establishing a space-based power grid to complement and interconnect power grids on the ground (Komerath 2011, Space power grid, personal communication to the author, 19 May 2011).

In the paper, the group anticipates that building an orbiting solar power production and exchange system will enable "a real-time power exchange through space to help locate new plants at ideal but remote sites, smooth supply fluctuations, reach high-valued markets, and achieve baseload status.

"Demand for power can vary with the time of day or the season of the year. The amount of electricity consumed on a hot summer evening can be 2–3 times greater than the amount consumed in the middle of the night during temperate weather. Because wind and terrestrial solar power sources are intermittent, auxiliary generators, which are expensive and fossil-burning, are needed at these plants to guarantee a steady baseload power flow. One advantage of nuclear power plants is that they can reliably meet baseload demand. Once these plants are up and running, they can be expected to supply consistent levels of energy 24/7, and they usually achieve nearly 90% or more of their rated power output year-round, compared to only 30–50% for wind and solar power farms" (Dessanti et al. 2011).

The Georgia Institute of Technology proposal takes a 50-year perspective, foreseeing a constellation of power-generating satellites capable of converting sunlight into as much as 4 terawatts of usable energy. This energy will be beamed to widely dispersed wholesale and retail markets on the ground. The first step toward this type of space power grid, according to the team, is a U.S.-India space-based power exchange demonstration that provides baseload energy across national boundaries. Two possible approaches to the first constellation achieving a near-24-h power exchange demo across countries are (1) four to six satellites at 5,500 km near-equatorial orbits, with ground stations in the United States, India, Australia and Egypt and (2) six satellites in 5,500 km orbits, with ground stations only in the United States and India.

"We argue for a strategy where SSP helps, rather than competes with, terrestrial renewable energy initiatives, as a way to establish the technology and the infrastructure to exchange power between markets."

"In other words, space is a venue for power exchange rather than just generation, and as such we call our architecture the Space Power Grid (SPG). This approach

Fig. 5.1 Artistic conception of a network of reflector satellites in equatorial orbit, relaying energy from a power sats in a Sun-synchronous orbit (Lightbourne 2011)

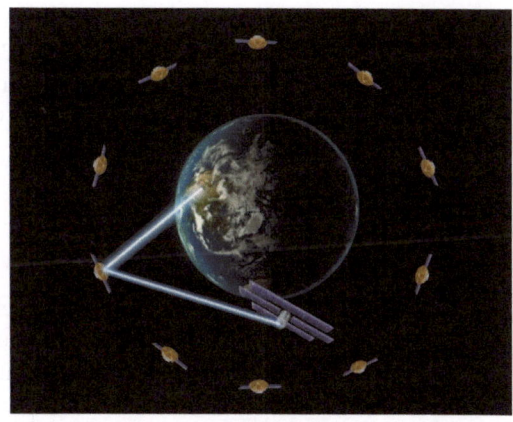

will also buy time to develop the best technological options for the gw-level SSP satellites that will replace the first-generation relay satellites. [Such a strategy] can lead to an economically viable infrastructure with a continuing revenue stream. This will help develop the massive satellites needed to expand SSP to the 4 terawatt level of today's fossil-based primary power supply" (Dessanti et al. 2011) (Fig. 5.1).

Historical Perspective

The scale and the potential impact of solar power satellite designs are much greater in 2011 than they were when the U.S. House Committee on Science and Technology asked for a study of the concept in 1978.

Responding to interest by NASA and the newly created Department of Energy, the House Committee sent a letter to the U.S. Office of Technology Assessment (OTA), asking that it undertake an independent look at "the potential of the SPS system as an alternative source of energy" and to assess its benefits and drawbacks as an energy system (Gibbons 1981, p. 18).

That report, entitled "Solar Power Satellites," was filed in August 1981. John Gibbons served as chair of the SPS Study Committee consisting of a distinguished advisory panel that included Peter G. Glaser, the widely acknowledged author of the concept.

The OTA Study Committee spent more than 2 years evaluating the prospects for solar power satellites. In a final summary, it wrote, "Along with other electric-generating technologies, SPS has the potential to supply several hundred gigawatts of baseload electrical power to the U.S. grid by the mid-twenty-first century. However, the ultimate need for SPS and its rate of development will depend on the rate of increase in demand for electricity, and the ability of other energy supply options to meet ultimate demand more competitively. SPS would be needed most if coal and/or conventional nuclear options are constrained and if demand for electricity is high."

The OTA study concluded, "SPS has potential for supplying a portion of U.S. electrical needs, but current knowledge about SPS, whether technical, environmental, or sociopolitical is still too tentative or uncertain to decide whether SPS would be a wise investment of the nation's resources" (Gibbons 1981, p. 55).

Public Policy Concerns

In the process of carrying out its research and deliberations, the OTA conducted an assessment of the potential environmental and human impacts of solar power satellites. This was perhaps the most thorough examination of such public policy issues as environment and health risks, land-use and receiver siting and military implications ever to have been done. It is because so many of the issues raised by the OTA study are the big "social impact" issues of today—many of which have yet to be fully addressed—that the author has chosen to highlight and quote at length from this 30-year-old source.[1]

Environment and Health

The OTA study reported, "Many of the environmental impacts associated with SPS are comparable in nature and magnitude to those resulting from other large-scale terrestrial energy technologies. A possible exception is coal, particularly if CO_2 concerns are proven justified. While these effects have not been quantified adequately, it is thought that conventional corrective measures could be prescribed to minimize their impacts" (Gibbons 1981, p. 10).

The study identified several health and environmental effects thought to be unique to SPS but whose severity and likelihood were uncertain. These included effects on the upper atmosphere from launch effluents and power transmission, human health hazards associated with non-ionizing radiation, radiation exposure for space workers and electromagnetic interference with other systems and with astronomy.

The authors understood that more research in these areas was needed before decisions about the deployment or development of SPS could be made, noting, "Little information is currently available on the environmental impacts of SPS designs other than the reference system. Clearly, environmental assessments of the alternative systems will be needed if choices are to be made between SPS designs" (Gibbons 1981, p. 11). "Reference system" refers to NASA's solar power satellite designs developed prior to the OTA study.

---

[1] In this section on early social concerns, the author trusts readers will see the value of an approach in which more questions are raised than solutions given when first introducing new and untested technologies. Current day research and expert opinion on most of these topics are addressed in Chap. 9.

The study team acknowledged that too little was known about the biological effects of long-term exposure to low-level microwave radiation to assess the health risks associated with SPS microwave systems. "The information that is available is incomplete and not directly relevant to SPS. Further research is critically needed in order to set human-health exposure limits. Currently, no microwave population exposure standard exists in the United States" (Gibbons 1981, p. 11).

The report continues, "More stringent microwave standards could increase land requirements and system cost or alter system design and feasibility. In light of the widespread proliferation of electromagnetic devices and the current controversy surrounding the use of microwave technologies, it is clear that increased under-standing of the effects of microwaves on living things is vitally needed even if SPS is never deployed."

There was concern related to the bio-effects of exposure to SPS power transmission and high voltage transmission lines on humans, animals and plants. "While the thermal effects of microwave radiation (i.e., heating) are well understood, research is critically needed to study the consequences of chronic exposure to low-level microwaves such as might be experienced by workers or the public outside of the receiver site."

"For SPS systems other than the microwave designs," the study leaders observed, "very little assessment of the health and safety effects has been conducted. The power density of a focused laser system beam could be sufficiently great to incinerate some biological matter. Outside the beam, scattered laser light could constitute an ocular and skin hazard." More study would be needed to quantify risks, define possible safety measures and explore the effects of long-term exposure to low-level laser light.

> The light delivered to Earth by the mirror system, even in combination with the ambient daylight, would never exceed that in the desert at high noon. The health impacts that might be adverse include psychological and physiological effects of 24-h-per-day sunlight and possible ocular damage from viewing the mirrors, especially through binoculars (Gibbons 1981, pp. 45–46).

Upper Atmosphere Effects

"Atmospheric effects result from two sources: heating by the power transmission beam and the emission of launch vehicle effluents. While the most significant effect of the laser and mirror systems is probably weather modification due to tropospheric heating, ionospheric heating is most important for the microwave systems operating at 2.45 GHz. Of most concern is disruption of telecommunications and surveillance systems from perturbations of the ionosphere" (Gibbons 1981, p. 45). The report explains further:

> Experiments indicate that the effects on telecommunications of heating the lower iono-sphere are negligible for the systems tested.
>
> The injection of rocket exhaust, particularly water vapor, into the ionosphere could lead to the depletion of large areas of the ionosphere. These "ionospheric holes" could degrade telecommunications systems that rely on the ionosphere. While the uncertainties are greatest for the lower ionosphere, experiments are needed to test more adequately telecommunications impacts and to improve our theoretical understanding of chemical-electrical interactions throughout the ionosphere.

In the troposphere, ground clouds generated during liftoff could modify local weather and air quality on a short-term basis. Additional experiments and improved atmospheric theory are needed to understand and quantify the above impacts under SPS conditions. In addition, mitigating steps such as trajectory control, alternate space vehicle design, and the mining of lunar materials need to be assessed. Atmospheric studies would play a major role in the choice of frequency for power transmission (Gibbons 1981, p. 45).

Land Use

The OTA study noted, "Receiver siting could be a major issue for each of the land-based SPS systems. Offshore siting and multiple use siting might each alleviate some of the difficulties associated with dedicated land-based receivers, but require further study. There are two components to the siting issue: technical and political. Tradeoffs must be made between a number of technical criteria:

- Finding geographically and meteorologically suitable areas.
- Finding sparsely populated areas.
- Keeping down the cost of power transmission lines and transportation to the construction site.
- Siting as close to the equator as possible (for GEO systems) so as to keep the north–south dimension of the receiver reasonably small.
- Coordinating receiver sites with utility grids and the regional need for electricity.
- The cost of land.
- Ensuring that the receivers are sited away from critical and sensitive facilities that might suffer from electromagnetic interference from SPS, e.g., military, communications, and nuclear power installations" (Gibbons 1981, p. 46).

It is clear that the choice of frequency, ionospheric heating limits, and radiation standards could have an impact on the land requirements. Further study is needed to understand fully the environmental and economic impacts of a receiver system on candidate sites and to determine if enough sites can be located to satisfy the technical requirements (Gibbons 1981, p. 46).

The earlier NASA technical (reference) designs had suggested the need for large contiguous plots of land dedicated to one use. The study's authors note that laser options might require less land area per site, but a greater number of sites to deliver the comparable amount of power.

The plausibility of multiple uses (e.g., agriculture or aquaculture), offshore siting (especially for such land-scarce areas as the northeastern United States, Europe and Japan) and possible receiver siting in other nations, with their particular environmental constraints, also need to be explored.

The report concluded that the regional political problems may be more severe than the technical ones, especially in light of past controversies over the siting of power plants, power lines, and military radar and other facilities. Although the construction and operation of receivers might be welcomed by some communities on the basis of economic benefit, others might oppose nearby receiver siting for a number of reasons, including: environmental, health and safety risks; fear that the receiver would be a target for nuclear attack; fear of decreased land values; preference for an alternate use of the land; objection to the receiver's visibility; and, for rural Americans, resistance to the intrusion of urban life (Gibbons 1981, p. 46).

Space Communications

An assumption of the writers of the OTA report was that all artificial Earth satellites would be using some portion of the electromagnetic spectrum for communication. Some would also use spectrum for remote sensing. All would be affected in one way or another by SPS (Gibbons 1981, p. 48).

Study members thought that geosynchronous satellites would be most strongly affected by the microwave systems, experiencing interference from noise at the 2.45 GHz frequency suggested in the reference design. "All radio frequency transmitters generate such noise and receivers are designed to filter out unwanted effects. However, the magnitude of the power level at the central frequency and in harmonic frequencies for a microwave SPS is so great that the possibility of degrading the performance of satellite receivers and transmitters from these spurious effects is high." The study continues:

> In addition to the direct effects from microwave power transmissions, geosynchronous satellites could also experience "multipath interference" from geostationary power satellites due to their sheer size. In this effect, microwave signals traveling in a straight line between (GEO) communications satellites would experience interference from the same signal reflected from the surface of the power satellite.
>
> The sum of all these effects would result in a limit on the distance that a geosynchronous satellite must have from the SPS in order to operate effectively. The minimum necessary spacing would depend directly on the physical design of the satellite, the wavelength at which it operated and the type of transmission device used (i.e., klystron, magnetron, solid-state device).

The study acknowledges that "There are numerous military and civilian satellites in various low-Earth orbits that might pass through an SPS microwave beam. Such satellites could in principle protect themselves from adverse interference from the SPS beam by shutting down uplink communications for that period, and improving shielding for data and attitude sensors, computer modules, and control functions" (Gibbons 1981, p. 50). The laser and mirror systems might also interfere with non-geosynchronous satellites by causing reflected sunlight to blind their optical sensors or by passing through communications beams.

Concluding Thoughts

The size of space/Earth antennas will certainly be a point of comparative difference between Sunsats and comsats, and so will the power levels of their transmissions to Earth. The footprints of early communications satellites—the spot on Earth illuminated by its power beams—were often as wide as one-third of Earth. In the case of today's comsats, their power beams are shaped so that the footprint conforms to specified coverage areas. Using spot beam technologies, such satellites can target areas of 100 square miles or less.

An estimated 300 currently active comsats are positioned in geosynchronous orbit (GEO). An even larger number of communications satellites are in MEO and LEO, including those collecting and using power for remote sensing, surveillance, weather, geo-positioning, satphone and military applications. According to a NASA website, that number might be as high as 3,100. Although their power ratings may be somewhat less, the total energy gathered and transmitted to Earth as microwaves is likely to be 10 times greater than those in the higher fixed orbit.

Orbiting comsats obviously collect and transmit less energy than is proposed for the new Sunsats. While the antennas of communications satellites are measured in meters and millimeters, those of solar power satellites will be measured in kilometers. Sunsat antennas will be sized to correspond to the total amount of the Sun's energy collected in space in ratio to the amount of usable energy needed for a specific purpose on the ground; thus, its receiving stations will be scaled to fit the need.

For siting and permitting, the U.S. government may have made Sunsat rectenna placement easier when it announced in late 2010 that it had established "solar energy zones" on public land in six western states and that other sites were under consideration. Large-scale solar energy projects within these zones were to receive streamlined authorization and preferential treatment. The announcement followed a report by the Departments of Interior and Energy of a 2-year environmental analysis of millions of acres of public land assessing environmental and other impacts of solar energy development.

"We think it provides a common-sense and flexible framework through which to grow our nation's renewable energy economy," Interior Secretary Ken Salazar said in a prepared statement. "Our country has incredible renewable resources, innovative entrepreneurs, a skilled workforce, and manufacturing know-how," Energy Secretary Steven Chu was quoted as saying, "It's time to harness these resources and lead in the global clean energy economy" (Environment 2010). Sunsat providers, in partnership with terrestrial solar businesses, may find future rectenna siting, and health, environmental and other public concerns easier to address as nations take steps to create more of their own energy.

References

Dessanti, B. et al. 2011. A U. S.-India power exchange towards a space power grid. Presentation given at the International Space Development Conference, Huntsville Ala., May 20, 2011.

Environment News Service. 2010. Solar energy zones identified in six western states. http://www.alternet.org/environment. Accessed 17 December 2010.

Gibbons, J. 1981. Solar power satellites. U.S. Office of Technology assessment. http://www.nss.org/settlement/ssp/library/1981-OTA-SolarPowerSatellites.pdf. Accessed.

Lore Lightborne Lore. 2011. Created in conjunction with Ohio University's International SunSat Design Competition. Based on an original concept by Exploration Partners, LLC. http://sunsat.gridlab.ohio.edu. Lightbourne Lore is a games and animation company founded by Digital Media students of Ohio University, Athens Ohio.

Chapter 6
What Is the Economic Basis for Solar Power Satellites?

Abstract This chapter addresses the financial attractiveness of clean and abundant energy delivered continuously to Earth 24-hours-a-day when compared with the rising energy costs and environmental damage caused by carbon-based energy sources. Steps forward are considered.

The Case for Sunsats

In 1995 Ralph Nansen wrote, "The future of mankind is dependent on abundant, low-cost energy that will not destroy the world." In his book *Sun Power: The Global Solution for the Coming Energy Crisis*, he asserts there is only one known source for that energy, and it is "giant [solar power] satellites [that will sit] in the silence of space, covered in a mantle of silky black solar cells, intercepting the life-giving rays and sending the energy to the Earth" (Nansen 1995, p. 6).

Nansen should know; he spent 31 years with the Boeing Company, primarily working in space engineering. As Boeing solar power satellite program manager, Nansen gathered the team of engineers, scientists and associated contractors that developed the overall SPS concept under the auspices of the Department of Energy and NASA.

In *Sun Power*, he identified five criteria by which any new energy source should be judged. It should be (1) non-depletable, sustainable; (2) non-polluting, environmentally clean; (3) low-cost, over a long period of time; (4) in usable form; and (5) be available to all (Nansen 1995, pp. 6–7).

Nansen also developed a series of interrelated rationales as to why the United States should commit to solar power satellite development, believing such an effort would:

- Give us a national purpose.
- Help us maintain our competitive edge in the world economy.
- Utilize the talents of our scientists, engineers and companies.

D.M. Flournoy, *Solar Power Satellites*, SpringerBriefs in Space Development, DOI 10.1007/978-1-4614-2000-2_6, © Don M. Flournoy 2012

- Free us from dependence on foreign oil.
- Enable us to better protect our environment.
- Open the space frontier for commercial development (Nansen 1995, p. 140).

Nansen had a unique opportunity to make his case before the U.S. House of Representatives Subcommittee on Space and Aeronautics in September 2000 when he testified on "the feasibility of space solar power." He had addressed the subcommittee in 1978 when he accompanied the president of Boeing Aerospace to testify on the same topic. This time he was speaking as the president of Solar Space Industries, a company he formed in 1993 to promote solar power satellite development.

"Much has changed in the last 22 years since I was here," he told the subcommittee, "but one thing that hasn't changed is the fact that solar power satellites are still not under development. However the time is now right for their development to begin" (Nansen 2009, p. 106).

Nansen went on to explain, "The studies conducted in the late 1970s determined the technical feasibility and the potential promise of solar power satellites for delivering abundant, low-cost, non-polluting electric energy to all the nations of the world. Studies since that time have reaffirmed this conclusion. In addition, much of the infrastructure that did not exist in the 1970s has been developed for other programs, dramatically reducing the development costs" (Nansen 2009, p. 106).

Nansen pointed out the plausibility of transmitting energy from one region on Earth that has excess energy capabilities to other world locations by reflecting wireless power transmission beams via relay satellites in space orbit. Because the relay satellites would be lightweight, they could more efficiently and economically be launched into space.

One of the key issues is what the government should be doing, Nansen told the subcommittee in 2000. His personal view was that development of solar power satellites should be primarily a commercial enterprise, but because of the size of such a program and its international implications, it should start as a government/industry partnership. "The primary role of the government would be to provide leadership and seed money to initiate the program, coordinate international agreements, support the development of high technology multi-use infrastructure, establish tax and funding incentives, and assume the risk of buying the first operational satellite" (Nansen 2009, p. 107).

Nansen was confident the energy produced by solar power satellites would create a large enough market if the perceived risk of their commercial viability were reduced to an acceptable level for the investment community.

Bilateral Project Development

In 2010, this author, as editor of the *Online Journal of Space Communication*, wrote an editorial in *Space News* lauding President Barak Obama's new National Space Policy, which supported "a robust and competitive space sector" (Flournoy et al. 2010a, b). This editorial was endorsed by the leadership of the Society of Satellite

Professionals International, the National Space Society and the for-profit Space Energy AG.

Among the goals of the President's National Space Policy was increased international cooperation on mutually beneficial space activities to "broaden and extend the benefits of space" and "further the peaceful use of space" (Obama 2010, p. 4) The editorial noted that these words represented "good news for those of us working to design and launch the new types of satellites that will collect solar energy in space and deliver it to Earth as a nonpolluting source of electrical power.... We believe space, as a shared resource, can best be explored and developed by a partnership of nations and businesses working together."

"Since acquiring clean and abundant energy is a common requirement for economic growth and an eventual necessity for the health of all societies, harvesting space solar power is a logical human endeavor when the high frontier is precisely where energy is most plentiful. But achieving success with large-scale commercial innovation in outer space requires long-range planning, pooling of financial resources, sharing of knowledge and expertise, and the careful framing of a way forward that will earn and sustain the public trust" (Flournoy et al. 2010a, b).

In naming the CEOs who would serve on his new advisory board on trade issues, President Obama noted in July 2010 that the United States is on track to double exports in the next 5 years, and he pointed to some of the ways the American economy is being repositioned to better compete abroad. When adding that announcement to the outcomes of the June 2010 Canada summit of the Group of 20 major industrial countries and recent federal policy statements intimating that certain export controls will be relaxed and cooperation in space will be encouraged, the signers of the *Space News* piece were hopeful that the United States would be entering a new era of openness for international business.

To this end, those who signed the editorial agreed, "We would like to see some greater leadership and support given to space solar power development by NASA and the U.S. Departments of Energy and Commerce. A helpful first step would be a U.S.-led space solar power feasibility study to which all interested nations are invited to contribute" (Flournoy et al. 2010a, b).

In the context of the new U.S. National Space Policy, the authors believed that a feasibility study could lead the way in assessing and promoting "appropriate cost and risk sharing among participating nations in international partnerships." Such a study would demonstrate the U.S.'s "tangible leadership in space," leveraging the capabilities of allies while assuring continuing adherence to the U.N. Treaty on Exploration and Use of Outer Space—now signed by 125 countries, including China and India—that dictates "nuclear weapons and other weapons of mass destruction" shall not be placed in outer space (Flournoy et al. 2010a, b).

The editorial noted that, at the International Space Development Conference held in Chicago in May 2010, multiple nations participated in a National Space Society-initiated Solar Power Symposium to examine in-depth the opportunities and challenges for energy generation in near space. Former Indian President A. P. J. Abdul Kalam, scientist, aeronautical engineer and proponent of space solar power, addressing the symposium via videoconference, spoke to the need for international cooperation in space, proposing a multilateral global initiative that could map out what needs to be done to bring space solar power into being.

"From our perspective," the editorial stated, "space solar power is a meaningful science, engineering and commercial challenge that deserves our attention and investment. In the wake of the Gulf of Mexico oil disaster, we think it is time for the U.S. to put space solar power on our national energy agenda. At the same time, we must seek opportunities to learn from and participate with Canada, China, India, Japan, the European Union and others taking their first tentative steps to bring space solar energy to Earth" (Flournoy et al. 2010a, b).

The editorial noted that in a June 2010 *Times of India* commentary on strategic international diplomacy, U.S. Senator John Kerry expressed support for a partnership with India that would include "the quest for new technologies and fresh ideas for economically viable ways to speed the shift to renewable energy sources."

The authors stated, "We believe that within the mainstream of global science, engineering and environmental management there are game-changing ideas and technologies that await testing. It is time to see some space solar power demonstration projects. Of all the possible alternative energy sources on the near horizon, we believe space solar power is our best chance for addressing the worldwide challenges of climate change, renewable energy and continued economic growth" (Flournoy et al. 2010a, b).

Indo-U.S. Collaboration

In April 2011, Rajeswari Pillai Rajagopalan, an analyst with India's Institute of Security Studies, wrote an article urging the United States and India to jointly develop an alternative energy source that would help the world free itself of nuclear technology, stating, "With the earthquake and the subsequent tsunami that hit Japan on March 11, isn't it time for India and the United States to make serious commitments to space-based solar power?" (Rajagopalan 2011).

She continues, "The Japanese crisis has triggered worldwide re-thinking on the feasibility of pursuing nuclear energy to meet growing global energy demands. This has kick-started a debate also in India not only on the safety of nuclear plants but also on other energy options. It is time that India and the United States and the countries around the world looked at an often-overlooked option: SBSP [space-based solar power]."

Dr. Rajagopalan pointed out that former Indian President Abdul Kalam had been a promoter of space solar power at the Aeronautical Society of India and more recently participated in a 2010 press conference on this topic hosted by the National Space Society in Washington, DC. The initiative to restart serious discussion of SBSP in conjunction with the U.S.-based National Space Society is now called the Kalam-NSS Energy Initiative.

She quoted from a speech Dr. Kalam gave in New Delhi in November 2010, writing that "by 2050, even if we use every available energy resource we have, clean and dirty, conventional and alternative, solar, wind, geothermal, nuclear, coal, oil, and gas, the world will fall short of the energy we need by 66%. There is an answer, an answer for both the developed and developing countries. This is a solar energy source that is close to infinite, an energy source that produces no carbon emissions, an energy source that can reach the most distant villages of the world, and an energy source that can turn countries into net energy exporters" (Rajagopalan 2011).

Dr. Rajagopalan further noted that the International Energy Agency predicts the worldwide demand for primary energy will increase by 55% between 2005 and 2030—a 1.8% hike per year on average. For India, the demand is expected to be more than double by 2030, growing at 3.6% per year, and in light of those figures, Rajagopalan questioned why SBSP is not being pursued:

> With energy demand growing rapidly, the SBSP option offers huge opportunities. Such an option will also be reportedly a cleaner energy option. This option would also significantly augment India's capabilities in the space domain, which will have far-reaching positive spin-offs in the ever-changing security environment in Asia. This will bring the much-desired focus on the question of technology transfer between India and the United States, Japan and Israel.
> What has prevented the SBSP from becoming a real option? Is it the enormous cost involved in developing the option or is it an option that never got the popular attention due to the multiplicity of departments involved? Proponents argue that the cost of SBSP should not be compared to the direct costs involved. The cost-benefit analysis needs to be done on a different scale, including the direct and indirect cost of global warming and climate change. Otherwise, the costs of developing this technology may seem exorbitant (Rajagopalan 2011).

Her article cited a 2009 U.S.-India agreement to establish a science and technology board and an endowment to carry out S & T research. She said this could be an enabling vehicle "because this fund seeks to finance projects on a broad spectrum of issues of mutual benefit such as biotechnology, health and infectious diseases, advanced materials and nanotechnology science, clean energy technologies, climate science, basic space and atmospheric and Earth science among others." The S & T "Rupee funds" were established in the 1980s to encourage and fund bilateral projects.

"While this can potentially be an excellent case for public–private partnership, the initiative has to come from the government. India's foray into space and its space policies have had strong civilian and developmental roots and accordingly the government needs to place the SBSP within its overall national space policy. India's decision to pursue SBSP will have multiple impacts—clean energy, clean environment, advancement in the space arena with technology transfer as a given between India, the United States and Japan" (Rajagopalan 2011).

The Commercial Sector

The commercial Space Energy Group AG has focused its business plans on both space-based and terrestrial solar power production.

Its website states that the company is "committed to becoming a world leading social enterprise—an organization driven by an ambitious vision to use commerce and innovation to change the world for the better." The company maintains that affordable, reliable, safe, clean energy is the catalyst for change in commercial and social dynamics that no other product or industry can match. Under the "About/ Why Space Energy?" section, the company reports:

> It is an indisputable fact that global energy usage is at a record high and continuing to rise fast. Demand in several areas of the global economy is already outpacing supply. Traditional hydrocarbon energy reserves are depleting at an ever-increasing rate, and most experts agree that there is only enough proven uranium reserves to last one more generation. In addition, the use of hydrocarbon and nuclear fuels are widely acknowledged to be leading contributors to significant environmental and health problems (Space Energy 2011).

Space Energy points out, "As developing countries continue to grow and embark on major electrification efforts, energy shortages will become one of the most serious challenges facing governments this century. China and India alone will need to raise energy-generating capacity by a staggering 4–5 times over the next 20 years in order to meet demand—an equivalent of bringing on-line two large coal-fired power stations per week, every week."

> The risk of energy shortages could mean more than high prices. In the twentieth century, many wars were motivated in part by the need to secure future energy supplies—and, according to the U. S. Pentagon, the risk of such conflict remains high in the twenty-first century.
>
> Aside from averting conflict associated with resource wars, abundant clean energy has the potential to truly improve life around the world in many ways. Rural electrification can offer one of the fastest ways out of poverty for developing areas. It can ensure that food and medicines are preserved and made available where they are needed the most. It can provide power for water purification and desalination and light so that children can study and develop their potential (Space Energy 2011).

In its May 2011 Space Energy Progress Report, the company notes, "Contrary to the recent global economic situation, the solar industry is thriving. Solar panel costs in the United States dropped by 12.5% just in the first quarter of 2011 and demand for solar power is rising, driven by measures such as the California law that requires the state to obtain a third of its energy from renewable sources by 2020" (Space Energy 2011).

"Individual, institutional and corporate investors are increasingly recognizing the potential of this industry. The first quarter of 2011 saw over \$2.5bn USD invested by venture capitalists in the clean technology sector, with the majority of that money going to solar power. This is a 13% rise from the year before."

Intermediate Steps

Globus, Barau and Radu have proposed strategies leading to an "early profitable powersat." Since space solar power implementation suffers from extremely large dimensions driven by the size and weight of on-orbit microwave antennas—requiring large capital inputs and long development cycles—this team outlines the merits of small, single-launch powersats that are designed to address niche markets. They say infrared power beaming based on fiber lasers and very lightweight collection structures using thin-film solar cells are potential solutions for bringing closer to financial feasibility single-launch-to-orbit in space solar power deployments. The approach they suggest is to keep launch and in-orbit collection costs down, but also to also deliver power (even though still expensive) to those clients, such as the U.S. military and other off-the-grid operations, currently paying a premium for energy (Globus et al. 2011).

These and other solutions are explored in an extensive "financial and organizational analysis" in connection with an aerospace management project conducted at the Toulouse Business School, Toulouse, France (Xin et al. 2009).

Concluding Thoughts

A prime goal of the Society of Satellite Professionals International, the professional development association of the satellite and space industry, is to help in expanding the satellite services market.

SSPI Director of Development Louis Zacharilla, paraphrasing business guru Peter Drucker, wrote in *SatMagazine*: "Without markets, or with markets that are in decline, competition becomes a desperate, zero-sum game. With expanding markets, opportunities emerge, innovation persists and capital flows. Expanding markets are virtuous, and in their wake the satellite community becomes more secure and attracts needed talent" (Zacharilla 2010).

The energy market is not on the horizon for the satellite services industry—not domestically, not internationally, not next year and possibly not in the next decade. Nevertheless, the signs point in that direction, which leads the author to predict that Sunsats, gathering the Sun's energy in space and delivering it to Earth as electrical power, will eventually dominate all other satellite businesses, including the currently very profitable comsat business.

Given the dire need for alternative sources of clean and abundant energy to avert global catastrophe, it is not hard to think of Sunsats as the latest new impact technology, the breakthrough development that expands the market, the business innovation that lifts the prospects of all related businesses. Satellite manufacturing and launch services, for example, will benefit in the near term, but also profiting will be all types of spinoffs to satisfy long-term needs here on Earth and in space.

Rather than sending comsats to the dustbin of history, competitive adaptation to the needs of the future energy market will be the basis for satellite services renewal. It can be imagined that healthy Sunsat systems successfully serving global energy markets will also be a big step forward in the further commercialization of space.

References

Flournoy, D., R. Bell, M. Hopkins, S. Tennsel, & F. Hsu. 2010a. International cooperation in space: Why not space solar power? *Space Energy.* http://spaceenergy.com.

Flournoy, D., R. Bell, M. Hopkins, S. Tennsel, & F. Hsu. 2010b. Why not space solar power? *Space News.* http://spacenews.com/commentaries.

Globus, A., I. Bararu, & M. Radu. 2011. Towards an early profitable powersat, Part II. http://space.alglobus.net/papers/TowardsAnEarlyProfitablePowerSatPartII.pdf.

Nansen, R. 2009. *Energy crisis: Solution from space.* Ontario, Canada: Apogee Books.

Nansen, R. 1995. *Sun power: The global solution for the coming energy crisis.* Seattle WA: Ocean Press.

Obama, B. 2010. *National space policy.* The White House. http://www.whitehouse.gov/the-press-office/fact-sheet-national-space-policy. Accessed 28 June 2011.

Rajagopalan, R. P. 2011. Space-based solar power: Time to put it on the new U. S.-India S&T Endowment Fund? Observer Research Foundation. http://www.nss.org/orf-*02April2011*.

Space Energy Team. 2011. Space energy progress report. *Space Energy Newsletter May 2011.* http://www.spaceenergy.com.

Xin, S., E. Panier, C. Zund, and R. Gomez. 2009. Financial and organizational analysis for a space solar power system: A business plan to make space solar power a reality. Toulouse Business School, Toulouse, France. http://www.nss.org/settlement/ssp/library/index.htm.

Zacharilla, L. 2010. Beam. Supporting the future of the industry. *SatMagazine October 2010. http://www.*satmagazine.com.

Chapter 7
What Are the Legal Issues?

Abstract This chapter explains the extent to which Sunsats can be deployed under existing treaties and regulatory provisions at various levels of government, and the extent to which new policies and procedures must be negotiated. Issues related to export controls, assignment of orbital positions and frequencies, ownership and control of space assets, liability for damage in space and environmental protection are also addressed.

International Development Goals

The International Telecommunications Union (ITU), at its May 2005 World Telecommunication Development Conference in Hyderbad, India, set broad goals for public access to ICTs (information and communication technologies), hoping to reach more than half of the world's population by 2015.

In its 2010 midterm review of these Millennium Development Goals, the helpful role of communication satellites was prominently mentioned. "If satellites are taken into account, then practically the whole world is covered by broadcasting," the report said. "The number of households around the world with DTH dishes rose from 82 million in 2000 to 177 million in 2008" (Oberst 2010, p. 14).

In reporting on the ITU midterm review in the trade magazine *Via Satellite,* Gerry Oberst noted, "This is not the end of the story, however, because access or coverage is not the same as actually receiving broadcasting signals. In addition to low income, the current lack of broadcasting reception in developing countries arises from lack of electricity.... The ITU statistics show that about 79% of the world's households own a television set, but only 28% of households in Africa own a set. To increase that number, satellite services offer the possibility for most developing countries to ensure national broadcasting coverage. Nevertheless, there is that tricky problem of a lack of electricity" (Oberst 2010, p. 14).

When it comes to satellite coverage, whether for solar power or for communication, politics and government regulations can play a decisive role. Prominent and long-standing examples of political impediments are the export rules imposed by the United States on global trade in satellites and satellite-related equipment beginning in 1999. These have come to be known as the U. S. International Traffic in Arms Regulations (ITAR).

In *The Broadband Millennium*, this author writes:

[In 1999, the] U. S. Congress wrote into a defense authorization bill language that placed limitations on satellite exports largely aimed at tightening U. S. technology transfers to China and curbing Chinese espionage in sensitive American facilities. The restrictions required detailed technology transfer control plans for any satellite or satellite technology to be sold outside U. S. jurisdiction, whether to China, Russia, Canada, or any trading partner nation.

With export licensing authority shifting from the Commerce Department to the U. S. Department of State, commercial satellite transactions were treated in the same manner as munitions transactions. Approvals for previously routine commercial exports and technical exchanges experienced long delays. At the time, U. S. companies were supplying 76% of the world's commercial GEO spacecraft and 88% of the LEO satellites. A Satellite Industry Association study found that by 2001 the U. S. share of the global market for communication spacecraft and parts had fallen to 45%.

The U. S. war on terrorism, implementation of Homeland Security measures, and the greater scrutiny given to international trade has made matters much worse for the global satellite industry. A particularly low point occurred when the U. S. National Defense Authorization Act for 2004 included "Buy American" provisions that would require the Pentagon to buy only hardware constructed with components and machine tools built in the United States" (Flournoy 2004, pp. 251–252).

By the end of his second term, President George W. Bush directed changes that would clarify regulations governing the export of civil aircraft components and streamline the U. S. export approval process. One of the first items on the agenda of newly elected President Barak Obama was to launch a review of all export control policies and procedures.

A 2009 editorial in the trade journal *Aviation Week & Space Technology* demanded, "Every facet of the export control regime must be on the table. Both the climate and the timing are ripe for major change. The Secretaries of Defense, State and Commerce all acknowledge the need for updated controls, and Congress is more aware than ever of the importance of defense exports to U.S. security and its economy" (Editorial 2009, p. 66).

Addressing the U. S. National Space Symposium in Colorado Springs in April 2011, Lei Fanpei, vice president of China Aerospace Science and Technology Corp. (CAST), spoke to the political and legal constraints hindering international cooperation in space. He made a direct appeal to the U. S. government "to lift its ban on most forms of U. S.-Chinese cooperation," saying both nations would benefit from closer government and commercial space interaction (de Selding 2011, p. 8).

Lei Fanpei was quoted as saying, "China purchased more than $1 billion in U. S.-built satellites in the 1990s before the de facto ban went into effect in 1999. Since then, the U. S. International Traffic in Arms Regulations (ITAR) have made it impossible to export most satellite components, or full satellites, to China for launch on China's now successful line of Long March rockets." He noted that "Chinese vehicles launched more than 20 U. S.-built satellites in the 1990s" (de Selding 2011, p. 8).

The government official from CAST suggested three areas of possible cooperation that would serve the interests of the two nations. These included open commercial access of each nation to the other's capabilities in satellites and launch vehicles, manned space-flight and space science—particularly in deep space exploration—and such satellite applications as disaster monitoring and management (de Selding 2011, p. 8).

Space Law

When Kiantar Betancourt wrote "Legal Challenges Facing Solar Power Satellites" for the *Online Journal of Space Communication*, he was a third-year student at the University of Maryland School of Law, specializing in environmental and international law. He currently works at Enhesa, Inc., an international consulting group. Permission is given for the abbreviated reporting of his article below, which is available in its entirety at http://spacejournal.ohio.edu/issue16/hsu.html (Betancourt 2010, p. 2).

In his article, Betancourt describes the current system of international space law, explaining the specific ways international regulations could help to create a supportive environment for launching, maintaining and removing solar power satellites. He also offers suggestions for future improvements to this system:

> Solar power satellites automatically raise questions concerning the currently applicable international law, and which laws and processes may need to be in place to accommodate the special requirements of SunSats.
>
> These questions include coordination and registration of space objects, property rights in space, rights of private parties, liability for damage, and environmental protection. The general framework to answer these questions already exists, but further development will be needed. The United Nations Committee on the Peaceful Uses of Outer Space (COPUOS) has led the development of this legal framework. Presently there are three treaties relating to outer space significant to SBSP.

He writes that the first and most important of these is the Treaty on Principles Governing the Activities of States in the Exploration and Use of Outer Space (Outer Space Treaty). Second is the Convention on International Liability for Damage Caused by Space Objects (Liability Convention). Third is the Convention on Registration of Objects Launched into Outer Space (Registration Convention)" (Betancourt 2010, p. 2).

The Outer Space Treaty

According to Betancourt, the Outer Space Treaty has been accepted and ratified by over 100 countries including all current spacefaring nations. Ratified in 1967, this treaty created the fundamental base of outer space law under the idea that outer space is the common heritage of mankind. Thus, the exploration and use of outer space shall be free for exploration and use by all states. Article II states that outer space, including the Moon and other celestial bodies, is not subject to national appropriation by any means. Even for countries that currently lack the resources to reach outer space, the right of exploration and use remains available to them as they become capable of space exploration.

Under Article VII, though a state cannot claim ownership to outer space or any celestial bodies within, a state on whose registry launches an object into outer space retains jurisdiction and control over that object. The ownership of such objects in outer space is also not affected by their presence in outer space or by their return to Earth. Thus, countries or companies that launch satellites on their state's registry retain ownership of those satellites. If no such ownership interest existed, there would be no incentive to send a satellite into space that could be appropriated by another country or private party.

Betancourt explains that the Outer Space Treaty addresses actions taken by states. It does, however, contemplate the actions of private companies in two sections. First, in Article VI, parties to the treaty agree to bear international responsibility for their national activities in outer space, whether those activities are carried out by governmental agencies or by non-governmental entities. Second, Article IX requires states and their nationals to seek international consultation in circumstance that could cause harm to other states. Though space exploration in 1968 was dominated by states, the Outer Space Treaty still contemplated private companies joining the states in space travel.

> The Outer Space Treaty contains several other key provisions. Article V of the Outer Space Treaty specifically prohibits the placement of any objects in space carrying nuclear weapons or weapons of mass destruction. Further, testing of any military weapons is strictly forbidden. An example might be an attempt to transform a solar power satellite into a death ray using microwaves or laser beams. Such an action would be in strict violation of the Outer Space Treaty.

He notes that Article XII of the Outer Space Treaty requires that any station, installation or equipment on the moon, asteroid or other celestial body must be open to inspection on a basis of reciprocity. This provision, though limited to objects on celestial bodies, allows countries to ensure that others are within the terms of the treaty. The Outer Space Treaty answers questions concerning the right of private ownership and the role of private companies in outer space (Betancourt 2010, p. 3).

The Liability Convention

"Ratified in 1972," Betancourt notes, "the Liability Convention helped clarify the liability of states and private parties for damage in space. The guidelines, under Article II, that affirmed that launching states will be absolutely liable for damage caused by their space objects on the surface of Earth and to aircraft in flight have now been approved and ratified by 91 countries including all current spacefaring nations."

He notes that countries have to create their own laws regulating private companies to protect themselves in the case that a company causes damage. If such regulations are not created, it could discourage a country from allowing a private company to go to space for fear of international liability. For example the United States passed the Commercial Space Launch Act of 2004 granting the Federal Aviation Administration the authority to regulate commercial space flights with the interest of promoting private space development while shielding itself from liability. Prior to launching an object into space, a private company has to apply for a license from the FAA. The CSLA requires all license applicants to demonstrate financial responsibility through liability insurance or independent means. The U. S. requires evidence of insurance to compensate another party for damages or itself for losses stemming from an activity carried out under the license. If the damage exceeds $500 million, the United States will cover the remainder up to $1.5 billion but only 'to the extent provided in an appropriation [bill].' Thus, anything over $1.5 billion would need to be covered by the company. If not enough money is allocated in an appropriations bill the company will be liable for all damages" (Betancourt 2010, p. 4).

Japan has taken a similar approach, he writes, but its law seems friendlier to private companies. "As in the United States, private companies have to secure liability insurance for an amount determined by the government. Unlike the United States, the government average liability insurance requirement is around $200 million. More importantly, the Japanese government will cover any amount over the liability insurance without limit." He points out that, even though Japan protects itself from potential liability, its approach makes it easier for private companies to enter into space.

As for solar power satellites, Betancourt recommends that countries continue to develop laws encouraging commercial space companies, which can help reduce development costs while bringing fresh ideas to the marketplace. Countries could provide further incentive to develop SBSP applications by lowering or eliminating a company's liability in exchange for the company's help (Betancourt 2010, p. 5).

Betancourt recommends that the United Nations and member states work together to clarify more precisely the meaning of "fault" so that countries and companies can more easily predict their potential liability. Thus, the international contingent should continue to develop the framework used to determine liability for damages, possibly to include requiring countries to clean up or retrieve broken or decommissioned satellites—or face strict liability for the damages they cause—and improving dispute mechanisms between countries and penalty assessment on those refusing to pay proper judgments. Penalties for refusal to pay for damages could help ensure damage award compliance, motivating countries and companies to promote safe practices, while lowering the risk of catastrophic losses (Betancourt 2010, p. 6).

The Registration Convention

In his article, Betancourt describes the creation of the Registration Convention and its importance to the evolution of the Sunsat industry:

> As more satellites entered orbits around Earth, the United Nations and its members recognized the necessity of registering all space objects in a single registry to help prevent accidental collisions in space. Ratified in 1974 by 53 countries, including all current spacefaring

nations, the Registration Convention, under Article II, requires all countries to create and maintain a registry of all objects they or their nationals have launched into space. Article IV then requires countries to give this information to the United Nations, including the objects' orbital parameters, from which the United Nations builds its global registry. Countries can then consult with the registry to ensure future satellites will not interfere with current ones. Private companies seeking to send up a satellite are expected to consult with their country registries to ensure the vehicle is noted domestically and that that information is submitted to the United Nations.

Betancourt notes that as more satellites are sent into space a simple registry may not be sufficient. The international regime will likely need to develop a mechanism for space traffic control with the ability to track satellites in orbit and the authority to assign orbital slots equitably, while establishing transit corridors for new satellites to safely reach orbit. Without such, space travel could become more dangerous. An increase in the frequency of collisions could also add to the costs and threaten the security of solar power satellites (Betancourt 2010, p. 7) (Fig. 7.1).

Space Debris

Based on his research, Betancourt concludes that space debris is the largest environmental problem for the SPS industry. He explains, "There are over 19,000 pieces of trackable debris in Earth orbit; the number of un-trackable pieces is much higher. Collisions with even small [pieces of] orbital debris can cause catastrophic damage."

The global community has taken steps to deal with this growing problem, he says. The Inter-agency Space Debris Coordination Committee (IADC) is an international organization made up of all major spacefaring countries, responsible for proposing solutions and researching problems posed by space debris. It has created guidelines to help minimize debris-creating events and avoid debris-caused hazards. The guidelines are not binding; however, states can use these guidelines to formulate their own mitigation standards. The United States also has its own standards to control space debris, and these standards offer initial guidance, but further improvements will be needed to fully address this problem. He writes:

> The Orbital Debris Mitigation Standard Practices (Standard Practices) of the U. S. government incorporates guidelines offered by the IADC while adding its own provisions. Like the IADC guidelines, the Standard Practices seek to avoid releasing debris during normal operations, especially debris larger than 5 mm that will remain in orbit over 25 years. The Standard Practices also offer guidelines for post mission disposal of space structures including:
>
> • Atmospheric reentry: for objects in LEO, where atmospheric drag should limit the lifetime of the object to no longer than 25 years;
> • Maneuvering the device to a storage orbit: structures would be moved or have the capability of moving themselves to different "storage" orbital levels; or
> • Direct retrieval: retrieving and removing the structure from orbit after completion of its mission (Betancourt 2010, p. 9).

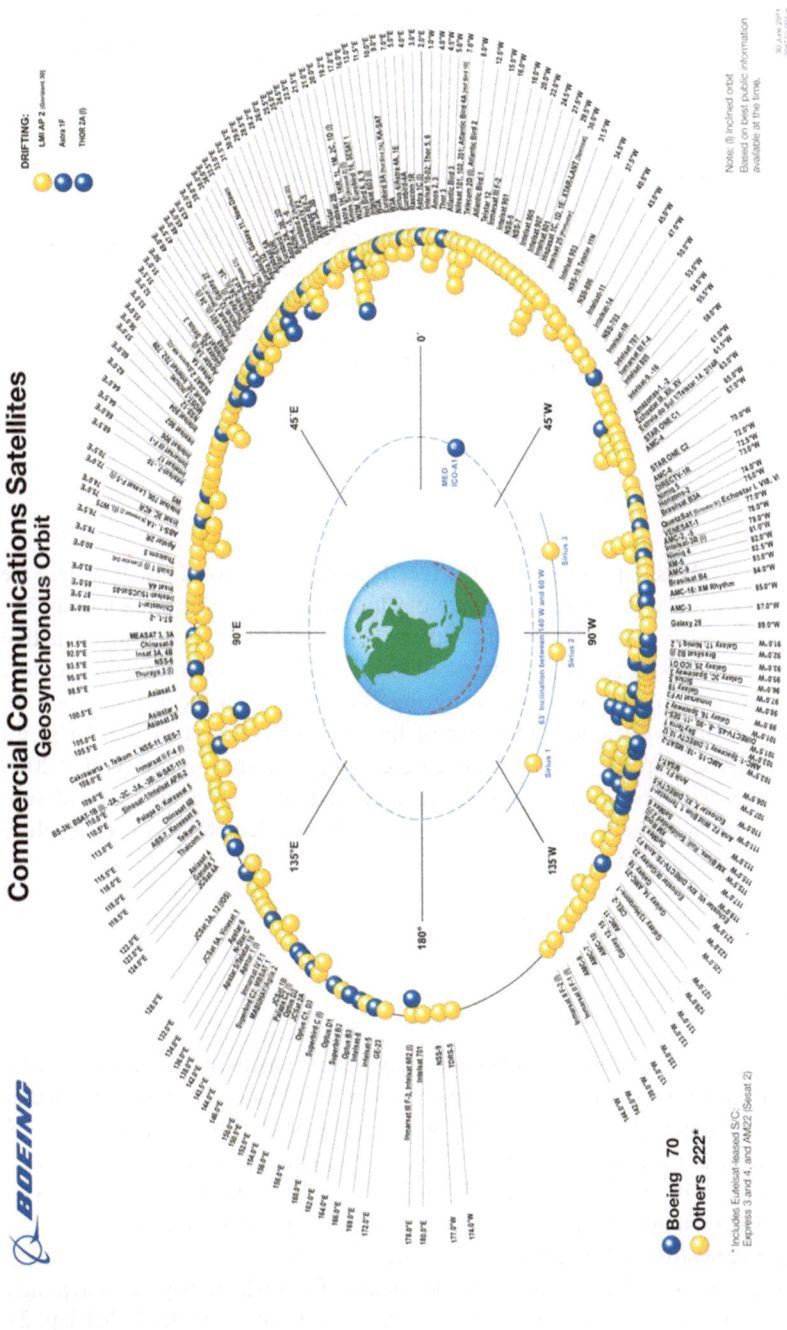

Fig. 7.1 Pictured are the commercial communication satellites within the 360° geosynchronous equatorial orbit that is 36,000 km above Earth. GEO is prime real estate in space since objects placed in this location move in sync with the turn of Earth (Boeing 2011)

Microwave Radiation

Betancourt's research led him to conclude that public health and safety issues with microwave use have been examined extensively. "Microwaves used in space solar power have no ionizing effect, and there is no danger of cancer or genetic alterations due to microwave radiation. The potential danger of microwaves, like energy from the Sun and from artificially light sources, relates directly to the energy's density in a given area. The design of SBSP systems calls for power densities well within safe limits at the planet's surface.

He explains, "For example, the average power density of the Sun's rays is about 100 mw/cm^2 while the design maximum of satellite solar power systems is 25 mw/cm^2 on the planet's surface." Even high-flying birds would still remain well within safe limits, he says, "though scientists should still plan further safety studies, a necessary precaution for technology on this scale" (Betancourt 2010, p. 10).

Other Regulatory Issues

Mark I. Wallach is a partner with Calfee, Halter & Griswold, LLP, where he serves as co-chair of the litigation group of more than 40 attorneys. A member of the National Space Society and the Space Frontier Foundation and an active advocate for space-based solar power, he contributed to the October 2007 *Report on Space Based Solar Power* issued by the National Space Security Office. In 2009, Mr. Wallach became a member of the Advisory Board of the for-profit Space Energy Group. He authored an article on legal issues in the winter 2010 *Online Journal of Space Communication*. Included below is a summary of several important matters he addresses that are likely to affect solar power satellite system implementation.

GEO Slot Rights

According to Wallach, a major, yet still largely undeveloped, legal question is who owns the right to the "slot" located at the geosynchronous orbit above a particular rectenna. He notes that "The highly prized equatorial orbit at approximately 36,000 km above mean sea level has the unique characteristic of appearing to maintain the same position relative to Earth's surface, since the object in that orbit has an orbital period matching Earth's rotational period. Ideally, SBSP satellites collecting energy and converting it into a microwave beam for transmission to the surface will be positioned in a suitable location over the equator, from which they can reach their targeted receiving rectennas by way of movable 'spot beams'" (Wallach 2010, p. 2).

Who owns—or who controls—the "air rights" in GEO orbit? Wallach cites the not-so-hypothetical example of a communications satellite already located there; does it have primacy by reason of prior arrival? If a company receives approval to locate its SBSP collecting satellite at a particular spot, is it entitled to that location in perpetuity, or for the life of the satellite? Wallach points out that, since most of the orbital slots in GEO have already been assigned to interested nations and not to individuals or companies, it will fall to the ITU and the nations' regulatory agencies to adjudicate such questions.

He explains that "The ITU, an agency of the United Nations, holds responsibility for assigning both orbital and electromagnetic spectrum positions. The ITU is governed by a constitution and the International Telecommunications Convention. The rights and obligations therein are binding on all member states. Currently, the ITU appears to apply a 'first-in-time, first-in-right' system to orbital allocation. However, the ITU's primary considerations are supposed to be equitable access and efficient use of a limited resource. Many argue that these considerations obligate the ITU to reserve spaces for developing nations."

> The matter of crowding is already a contentious issue for present and future operators of satellites at GEO. Telecommunications satellites need to be positioned far enough away from one another to ensure that their signals do not interfere with each other. The ITU Radio Communication Sector interprets, administers, and enforces the policies and agreements of the ITU, and also oversees coordination of the use of the spectrum and assists in solving conflicts with orbital position in its "Master Register" (Wallach 2010, p. 3)

Wallach notes that Article II of the Space Treaty assures that outer space "is not subject to national appropriation by claim of sovereignty, by means of use or occupation, or by any other means." The Space Treaty also appears to prevent private companies from selling slots in the geostationary orbit: "Under the current treaty regime, the geostationary orbit is a scarce resource that no nation or individual can claim a legal right to beyond that of a squatter, which does not work to allocate the orbital space either efficiently or equitably…. While the ITU presumably will govern the allocation of GEO slots to SBSP satellites, it is by no means clear how conflicts between communications satellites and their vastly larger SBSP cousins will be decided, or what criteria will be used to make those decisions" (Wallach 2010, pp. 4–6).

Power Beaming

Wallach cites another legal issue that relates to the operation of SBSP systems. That is, "Transmission of microwave beams to the ground may be subject to the jurisdiction of the Federal Communications Commission (FCC), which has asserted the right to regulate even very small-scale demonstrations of wireless power transfer. What degree of possible interference with other wireless power transfers—such as radio broadcast signals, cell phone communications, and television broadcasts—will or should be tolerated? What is the extent of FCC jurisdiction over an activity that is typically thought of as within the jurisdiction of state public utilities commission: supplying electric power?"

Certain federal regulations, specifically 47 C.F.R. §§ 101.4–101.97, govern the application and licensing of fixed microwave services. Likewise, 47 C.F.R. §§ 25.110–25.165 govern the application and licensing for all satellite communications. Under these regulations, the FCC is charged with granting such licenses. There are also temporary options during the pendency of licensing applications. For example, 47 C.F.R. § 101.31(b) grants applicants for new point-to-point microwave radio stations, or modifications of existing stations, authority to operate during the pending period of a licensing application if certain conditions are met. Thus, it seems that the FCC would also be responsible for the regulation of the SBSP microwave beam, and the granting of any necessary licenses (Wallach 2010, p. 6).

Wallach foresees the power beam itself raising regulatory questions. Even though the low intensity of the beams—which will spread out to an area of one square mile or more by the time they reach Earth's surface—ensures that they are not a health risk to humans, these beams could nevertheless affect, for example, the migratory pathways of birds. Is that an issue for state departments of natural resources, or some federal agency? He continues:

And what effect, if any, will the beams have on airplane traffic? Will no-fly zones be created over rectennas? Or simply some kind of warning signal for aircraft approaching the space over a rectenna? As for air traffic, probably such questions will have to be determined, at least in the first instance, by the Federal Aviation Administration (FAA).

He notes that an alternative method for transferring power from SBSP collector satellites to ground stations is with high-intensity laser beams, especially for smaller systems (because microwave power transfer systems do not scale down well). In that case, more serious safety issues could arise, including liability for property damage or even personal injury by diverted laser beams. Since low intensity microwave beams pose no health threats, personal injury liability is not a consideration; but the same cannot be said about high-intensity laser beams (Wallach 2010, pp. 7–10).

Renewable Energy Targets

"Perhaps the first issue raised by SBSP power contracts will be whether those contracts can be used to satisfy regulatory targets for renewable energy," Wallach says, and he goes on to note that, for conventional renewable sources, this question may be answered by the specifics of state regulatory requirements. But some states may insist that power actually be produced and purchased to meet renewable energy targets, while others may only require that those utilities have entered into good-faith contracts with providers of qualifying energy.

He notes that in California, for instance, public opinion holds that the PG&E/Solaren contract, approved by the California Public Utilities Commission in 2009, is useful whether or not it could be performed. "The law appears to be fairly stringent; that is, Section 399.15 of the California Public Utilities Code requires that the specified purchase levels be procured from eligible renewable energy resources" (Wallach 2010, p. 6).

Wallach consulted a report published by the California Energy Commission that discussed the risks of signed renewable energy contracts failing to meet the timelines in the contracts and found "this risk of contract failure could cause individual load-serving entities, or entire states, to fall short of their renewable energy targets." The report suggested that companies should anticipate a contract failure rate of 20–30%. This led to the conclusion that simply because a company has a contract in place to procure renewable energy, the contract will not, by itself, satisfy the regulation unless it is actually procured (Wallach 2010, p. 7).

The Role of Government

Feng Hsu, vice president for systems engineering and risk management at the Space Energy Group, is of the opinion that a model similar to the one used in successfully launching and commercializing communications satellites will be a viable approach for solar power satellite implementation.

As a former NASA scientist, serving as a senior advisor to the Aerospace Technology Working Group and a co-founder of the Space Development Steering Committee, Dr. Hsu has been an advocate for space-based solar power for a long time. In thinking about "the roadmap ahead," he believes that hope for a viable solar power satellite system lies in the collaborative efforts of private, entrepreneurial space businesses and venture capital investment, undertaken as a global-scale commercial enterprise.

He writes that "For SPS to be successful, we need an organized consortium consisting of private businesses, venture capitalists from major international partners, along with government support of R&D and technology demonstrations by industrial nations. We need this concerted effort to bring down associated risks in safety, reliability and technology maturity." He also says he is convinced that government policy and regulatory support will be crucial to success, as will the funding of R & D and related technology demonstrations, "but quite frankly, as a former employee of one of the great space agencies of the world, I am pessimistic about getting the necessary government support for any SBSP project" (Hsu 2010, p. 6).

Concluding Thoughts

Some of the legal and policy issues identified by Betancourt, Wallach and Hsu are unique to Sunsats and could require special attention, adjudication and perhaps some law-making. But their research and experience suggests that the preponderance of current regulatory concerns about solar power satellites have been anticipated in law and in regulation.

Doing the legal research and anticipating regulatory roadblocks are important and necessary steps to establishing the SPS industry. But even more important is realizing that those companies, those utilities, those nations aspiring to be in the business of providing energy from space are lucky to have a mature and profitable comsat industry at hand showing them the way, and that solar power satellites and communications satellites are natural allies. The author concludes: the one is the natural business extension of the other.

References

Betancourt, K. 2010. Legal challenges facing solar power satellites. *Online Journal of Space Communication*. http://spacejournal.ohio.edu/issue16/betancourt.html. Accessed 20 May 2011.
Boeing Company. 2011. http://www.boeing.com/defense-space/space/bss/launch/980031_001.pdf. 30 June 2011.
de Selding, P. B. 2011. Chinese government official urges U.S.-Chinese space cooperation. *Space News*. http://www.spacenews.com/civil/110414-chinese-official-space-cooperation.html. Accessed 18 April 2011.
Flournoy, D. 2004. The broadband millennium: Communication technologies and markets. Chicago: International Engineering Consortium.
Hsu, F. 2010. Harnessing the Sun: Embarking on humanity's next giant leap. *Online Journal of Space Communication*. http://spacejournal.ohio.edu/issue16/hsu.html. Accessed 20 May 2011.
Editorial. (2009). Modernize export controls. *Aviation Week & Space Technology*. 7 September 2009.
Oberst, G. 2010. Global regulations: ITU satellite goals. *Via Satellite*. September 2010.
Wallach, M. I. 2010. Legal issues for space based solar power. *Online Journal of Space Communication*. http://spacejournal.ohio.edu/issue16/wallach.html. Accessed 20 May 2011.

Chapter 8
How Is Sunsat Development Faring Internationally?

Abstract This chapter provides a brief summary of the types of research and development projects being undertaken by China, India and Japan, the three counties most likely to launch demonstration solar power satellite projects in the foreseeable future. Instances of cross-nation collaboration aimed at keeping their options open are also noted, including those of Europe and the United States.

SPS over China

The author of this book was co-chair of the International Conference on Space Information Technology in 2005, 2007, and 2009, hosted by the Chinese Academy of Space Technology (CAST). The first two of these were held at Huazhong University of Science and Technology, Wuhan, China. The third conference was held at Space City in Beijing, with the additional sponsorship of the China Aerospace Science and Industry Corporation, and several prominent Chinese and other universities. CAST is considered to be the Chinese near-equivalent to NASA.

Just prior to the November 2009 conference, where I was to present a keynote address entitled "Solar Power Satellites: Our Next Generation of Satellites Delivering the Sun's Energy to Earth" (Flournoy 2009), I was invited by the CAST Research and Development unit to present a "History of Space Solar Power R & D" at a special morning seminar. I agreed to attend the meeting and make a brief presentation but only on the condition that the session would be informal and interactive and include a report on the status of solar power satellite research and development in China, which at the time was not widely known (Fig. 8.1).

When I arrived at the seminar, I found present about 20 CAST senior and junior researchers—including the aging but still very engaged father of the Chinese space

Fig. 8.1 A 2009 meeting of the space solar power R & D group at the China Academy of Space Technology, including distinguished space scientist Wang Xiji (Photo by author)

industry, Wang Xiji[1]—with not one single other foreign professional in the room. Although the exchange was cordial and informative, the CAST staff reticence to talk about their own projects or their goals was apparent.

This meeting was something of a breakthrough, however, for learning about Chinese research and development regarding solar power satellites. Six months later, three of the senior scientists present at the Beijing seminar were able to obtain clearance for the first-time publication of the official Chinese vision, strategy and schedule for space-based solar production. Their paper now appears in the *Online Journal of Space Communication* as "Solar Power Satellites Research in China" (Gao et al. 2010). Readers can consult the original in its online version at http://spacejournal.ohio.edu/issue16/ji.html, but for the purpose of this book, I have

---

[1] Wang Xiji is advisor to the Chinese Academy of Space Technology. Born in Dali, Yunnan Province, he received his B. Eng. degree from National Southwest Associated University in 1942 and M.S. from Virginia Polytechnic Institute, U.S.A., in 1949. In the 1960s, Wang was in charge of the research and development of 12 types of sounding rockets and the technical design of China's first launch vehicle, Long March 1. He was the first chief designer of Chinese recoverable satellites and the first in China to propose the view that manned space technology must be developed to exploit space resources.

selected, summarized and highlighted below the items I thought most significant in the CAST report:

China's Long-Term Vision

So what is the official position of China on space-based solar power? The reader cannot mistake the country's near-term and long-term intentions.

> The state has decided that power coming from outside of Earth, such as solar power and development of other space energy resources, is to be China's future direction. The responsibility for ensuring China's food safety for its huge population, meeting its international obligations for environmental protection, and providing the structure for its energy needs have determined that the direction of future development of low-carbon energy sources cannot be to sacrifice the 'inner' Earth.
>
> Space-based solar power (SBSP), and the development of solar power satellites (SPS) to facilitate renewable energy production, is one of the "outside" approaches currently under development in China (Gao et al. 2010).

China's Energy Future

> In 2008, China's total energy consumption reached 2.85 billion tons of standard coal, while its electricity consumption reached 3.45 trillion kwh, a recorded 5.6% increase over the previous year. The annual report on China's energy development, pointing to the prospect for future energy demand, shows that in 2020, 2030 and 2050, China's total energy consumption of standard coal will climb to 3.5, 4.2, and 5.0 billion tons, respectively. In 2050, about 85% of the growth in energy demand can feed from fossil fuels, from nuclear power and from hydropower.
>
> Only 30% of the remaining 15% of that growth in energy demand can be met by non-hydro renewable energy resources, such as wind power, bio-energy, terrestrial solar power and tidal energy. That means that by 2050, despite China's continuing growth in energy production based on traditional energy areas, there is a considerable energy gap (approx. 10.5%), for which the state must look to such newer energy-producing approaches as fusion and space power stations.
>
> The Chinese Academy of Engineering's (CAE) cautionary report has shown that the fossil energy reserves in China, such as oil, coal and natural gas, will be exhausted in the next 15, 82 and 46 years correspondingly. How to fix the perceived loss of traditional energy resources has become an important problem for China's government. The CAE report also raises the question of growing public concerns over higher fossil fuel prices.
>
> In a 2009 global environmental summit in Copenhagen, the Chinese government promised that by 2020 China's greenhouse gas emissions will be reduced 40% compared with 2005. It suggests that the government believes that continuing to develop energy resources and environment protection are not internally inconsistent, and that low-carbon energy has a promising future in China (Gao et al. 2010).

Sustainable Development

China is thirsty for energy to water its blooming industries. SPS is regarded as a reasonable path to energy production. Either from GEO or LEO, this type of power system will have more direct access to the power of the Sun. In analyzing the characteristics of SPS and space solar power applications, CAST concludes that the advantages of SPS for China can be grouped into three relevant directions. [The first of these is] sustainable economic and social development.

With its population growth and rapid economic development, over the next 30 years China will become one of the most powerful and influential economies in the world. During this time, energy resources and environmental issues will be serious challenges for China. To avoid the grave consequences and to learn lessons drawn from others' mistakes, a sustainable development strategy will need to be adopted. This strategy can be expected to include renewable energy sources from outside Earth to alter the heavy reliance on fossil fuels, a process that will contribute to world energy development and assure environmental protection.

The acquisition of space solar power will require development of fundamental new aerospace technologies, such as revolutionary launch approaches, ultra-thin solar arrays, on-orbit manufacture/assembly/integration (MAI), precise attitude control, *in-situ* resource utilization (ISRU) for deep space exploration and colonial expansion into space. Since SPS development will be a huge project, it will be considered the equivalent of an Apollo program for energy. In the last century, America's leading position in science and technology worldwide was inextricably linked with technological advances associated with implementation of the Apollo program. Likewise, as China's current achievements in aerospace technology are built upon with its successive generations of satellite projects in space, China will use its capabilities in space science to assure sustainable development of energy from space (Gao et al. 2010).

A Skilled Workforce

Another priority in solar power satellite development, according the CAST research and development report, is in the area of "retaining and cultivating talent."

China understands that having an innovative, qualified and skilled workforce is the basic infrastructure on which national development can proceed. Higher education in China is developing rapidly, but the state lacks talent at both ends of its research lines, that is, in advanced concept research and in basic/technical sciences research. Objectively and actually, these are currently greater problems than finding financial sources for research.

CAST is of the opinion that in order to attract more outstanding personnel and to generate a magnetic field for attracting more college students into basic sciences and engineering, it is necessary for China to launch an SPS-type Apollo project to increase research and development investment in all corollary fields. This will relate to the country's goal of attaining the leading position in both energy and space technology (Gao et al. 2010).

Heading Off and Mitigating Disasters

The third priority for Chinese R & D in development of space solar power relates to disaster prevention and mitigation.

> In 2005, Hurricane Katrina killed thousands of people in the United States. Meanwhile, every year several typhoons bother the east coast of China. From preliminary research, it appears that microwave wireless power transmission may heat the top of the clouds, thereby reducing the force of typhoons and hurricanes.
>
> In 2008, China's southern region experienced a rare snowstorm; such extreme weather led to a complete paralysis of the entire southern power grid due to frozen equipment. Without wired power supplies, the economy of the southern provinces suffered heavy losses in the first few months of 2008. Again, if there had been an operational SPS power system in China, wireless power transmission quite possibly could have unfrozen the grid and restored power to the region.
>
> In May 2008, in the Sichuan region, a deadly earthquake measured at 8.0 magnitude killed thousands. The most important steps to be taken in mitigating the effects of that earthquake was to rebuild the human support system and provide an alternative communications system, each of which depended on the reinstatement of power supply systems. As space satellite systems can help to supply prompt restoration of terrestrial communications, and space solar power systems can achieve wireless power transmission via microwave and laser beams, space-based solutions would have been the fastest and most appropriate way to attack those problems (Gao et al. 2010).

SPS Implementation

The CAST SPS research team noted that there were four important areas of development: launching approach, in-orbit construction/multi-agents, high efficiency solar conversion and wireless transmission. "Except for launch," they asserted, "the other aspects do not seem to be insurmountable issues for China in the upcoming years."

> Based on China's 2010 space solar power plans, five steps would be taken in achieving its SPS system. In 2010, CAST will finish the concept design; in 2020, it will finish the industrial level testing of in-orbit construction and wireless transmissions. In 2025, it will complete the first 100 kw SPS demonstration at LEO; and in 2035, its 100 mw SPS will have electric generating capacity. Finally in 2050, the first commercial level SPS system will be in operation at GEO (Gao et al. 2010).

In August 2011, addressing a Beijing conference on Energy and the Environment, the 90-year old China space pioneer Wang Xiji updated that scenario. Speaking for CAST, Wang indicated that the detailed design of system solutions and key technologies as well as key technologies for authentication would be completed by 2020, and a space solar energy station for commercial use would be in service by 2040. Wang said he believes such a station will trigger a technical revolution in the fields of new energy, new material, solar power and electricity (China 2011). This announce-ment to accelerate the pace of space solar research and development came after 20 national academicians countersigned the China SPS Report and appealed directly to China Premier Wen Jiabao.

SPS over India

Raghavan Gopalaswami was chairman and managing director of the Indian aerospace company Bharat Dynamics Ltd. before retiring in 1994. He is noted for his pioneering research and design on rocket propulsion systems. From 2002 to 2007, he served as a non-official advisor to Indian President Dr. A. P. J. Abdul Kalam in formulating India's national Vision 2020, establishing goals and policies in the aerospace and renewable energy sectors of the national economy and security. This included giving advice on solar satellite and advanced space transportation systems, areas mentored by Dr. Abdul Kalam since their inception in India in 1987.

In 2010, Gopalaswami wrote an article for the *Online Journal of Space Communication* entitled "Sustaining India's Economic Growth" in which he exam-ined the energy policies of his country (Gopalaswami 2010). Excerpts from that article are provided below.

> India is among the few spacefaring nations of the world who have the capability to effectively participate in global missions for space solar power and related space transportation systems.
>
> Throughout the 1990s, advocacy for SSP increased in India and the United States. At an International Conference on High Speed Air & Space Transportation in Hyderabad in June 2007, organized by the Aeronautical and Astronautical Societies of India, leaders from the Defense Research and Development Organization and the Indian Space Research Organization advocated a global aerospace and energy mission. They placed on record their recommendation that "there is a need to generate a national consensus for the Global Aerospace & Energy Initiative, determine the sources and uses of funding, and evolve a suitable management structure and system to plan and implement the mission."
>
> India's interest in SSP originated in 1987, with the conceptual design of a single-stage-to-orbit fully reusable aerospace vehicle called Hyperplane. Over 22 reusable aero-space launch vehicles have since been designed the world over. None has been made operational so far. ISRO has taken up a program to develop an RLV (reusable launch vehi-cle) Technology Demonstrator somewhat on the lines of Japan's Hope concept.

Research in India showed the linkage between achieving high payload fraction from 10% to15% of takeoff mass. New space transportation systems concepts and technologies were presented for achieving high hydrogen fractions up to 60%. To achieve a cost of $100–200 per kg in LEO, an SSP transportation system has to be reusable at least 100 times, and have a payload fraction at least 10 times that of the shuttle, namely, about 15%.

Gopalaswami recommended that India prepare a Detailed Feasibility Study on Space Solar Power and a Reusable Space Transportation System as an integrated mission and systems design effort with assistance from other interested nations. This would include "advanced, reliable space transportation systems that have high payload efficiency [>10%] and payload delivery costs <$200/kg specifically for the SSP payload as well as Space Solar Power satellites and orbital assembly technologies."

A senior advisor from India estimates that the funding for a Detailed Feasibility Study and a critical technology demonstration could be around $200 million over 2–3 years (Gopalaswami 2010).

India's Energy Policies

Gopalaswami projects that a 30-year "business-as-usual" approach using coal-based thermal power plants, hydroelectric, wind and solar, with the addition of nuclear, would yield GDP growth rates in India ranging from 3.5% per annum to 5.5%. "Even with 5.5% GDP growth, the nation will have to increase annual power capacity from an historical (1950–2009) peak of about 4 gw/year to unprecedented levels of 18 gw/year in 2032 and 28 gw/year by 2052."

India's emphasis is now on terrestrial solar power, he said. Although terrestrial solar availability is limited to an average of 5.6 h/day, this type of energy has the advantage that it is clean and perennial. Solar energy harvested in space has the advantage that it is a 24/7 source capable of producing much greater quantities of energy. "For a sustained 7% GDP growth rate targeting 1,476 gw in 2052, and as an 'insurance policy' for shortfalls in achieving power capacity growth using terrestrial sources," space solar power could contribute an additional 17 gw in 2017 to 544 gw in 2052. This added SSP capacity almost doubles India's per capita GDP, delivering a net GDP benefit to the nation estimated to be worth over $100 trillion. The net carbon avoided by SSP substitution would be about 66 million tons" (Gopalaswami 2010).

India's Strategic Goals

Gopalaswami reminds his reader that India's population exceeds one billion, while its per capita GDP is among the lowest in the world. "India stands 134th in the Human Development Index among nations. Climate change is expected to have an adverse impact on economic growth among developing countries, especially in southern Asia. Energy is widely thought to be the principal engine for economic growth. Access to energy can multiply human labor and increase productivity in agriculture, in industry and in services. To sustain economic growth, energy supplies have to grow in tandem" (Gopalaswami 2010).

He notes that the Planning Commission of India in its Integrated Energy Policy Report of August 2006 advocated that an 8% GDP growth rate be sustained for the next 25 years. "In June 2008, the Prime Minister of India announced a National Solar Mission, recommending a massive build-up of terrestrially distributed solar power plants for rural areas. This mission has the potential to directly accelerate human development and reduce poverty levels in a manner that could not be achieved in centuries gone by. In May 2009, the government of India announced a targeted GDP growth rate of 7% per annum. In June 2009 the Prime Minister, addressing Parliament, urged the nation to target for 9% GDP growth rate" (Gopalaswami 2010).

Power Capacity Constraints

Gopalaswami details the circumstances that inhibit India's growth in terms of power.

> Foremost among the reasons for growth-limiting constraints on power capacity in India are: (1) political turmoil and serious breakdown of law and order in the land acquisition process, especially when diverting scarce agricultural land for industrial purposes; (2) tendering tangles, delayed statutory clearances even when acquiring non-arable land; (3) disjointed fuel supply chains and a severe shortage in facilities to manufacture power equipment (note that India has entered a phase where its industrial capabilities appear inadequate to expand electric power plants, a hen-and-egg situation); and (4) severe shortages in water supplies for these power plants, due to the drying and silting up of rivers and other water sources associated with climate change.
>
> Rapid capacity expansion is important for ambitious rural electrification. Power-for-all programs and shortfall in power capacity buildup is compelling states to give away thousands of crores (billions of dollars) worth of electricity free to the farming community for agricultural operations. For just 5.5% GDP growth, the nation has to gear up for annual power capacity additions from the historical peak of about 4 gw/year to unprecedented levels of 18 gw/year in 2032 to 28 gw/year by 2052. This growth must be achieved in the face of severe economic, environmental and other constraints (Gopalaswami 2010).

Former President A. P. J. Abdul Kalam has explained that his interest in space-based solar power came from the need to meet India's growing energy requirements while moving away from fossil fuels. He was quoted in the *Space Review* as saying, "We need to graduate from fossil fuels to renewable energy sources" (Foust 2010). In collaboration with the U.S.-based National Space Society, Dr. Kalam has proposed that an International Space Solar Power Feasibility Study be undertaken as a precursor to SSP/RLV technology demonstrations on the ground and in space.

President Kalam was an invited speaker at the August 2011 China Energy and Environment Summit, addressing participants via teleconference. He spoke of his own experience as a space scientist and the space vision required to achieve societal missions from space, which necessitates "advocating International cooperation for the large scale space missions including space based solar satellites." The question is, he said, whether we can "graduate in the ensuing years to partnership missions among space faring nations for the benefit of entire humanity" using the core competence of multiple nations and financial sharing (Kalam 2011).

He advocates for a new World Space Vision 2050 with three components: (1) Large scale societal missions and low cost access to space in which space faring nations work together; (2) Comprehensive space security in which all space faring nations participate and contribute to protect world space assets; and (3) Space exploration and application missions in which Earth-Moon-Mars are a single economic complex for the benefit of humanity. Such a vision would enhance the quality of human life, inspire the spirit of space exploration, expand the horizons of knowledge, and ensure space security for all nations of the world.

To realize these missions, President Kalam would like to see a "World Knowledge Platform" in which we share knowledge worldwide, freely exchanging data and information to establish the technical and economic feasibility of designing, building and operating a system-of-systems, consisting of low cost space transportation and space solar satellites. Once a feasibility study—that would include technology demonstrations of SSP and its critical enabling technologies—is completed, then action could commence on commercial scale implementation, bringing new clean green energy from space for rejuvenating our planet (Kalam 2011).

Peter Garretson, a U. S. Air Force lieutenant colonel, spent a year at the India Institute for Defense Studies and Analyses, developing "a policymaker's overview" of a potential Indo-U. S. strategic partnership in the context of space solar power.

Garretson's report concludes "that SBSP does appear to be a good fit for the U. S. domestic, Indian domestic and bilateral agendas, and there is adequate political space and precursor agreements to begin a bilateral program, should policymakers desire it. Given that SBSP appears to fit the articulated Indian criteria for suitability of energy source and to offer a better long-term energy security solution, and that the evaluation of the current energy-climate situation is so unhopeful, with a lack of promising and scalable solutions emerging, a no-regret, due-diligence effort in space-based solar power seems a justified and strategic investment" (Garretson 2010).

Garretson continues, "An actionable, three-tiered program, with threshold criteria/ goals, has been proposed, moving from basic technology and capacity building to a multi-lateral demo, and ultimately to an international commercial public-private partnership entity to supply commercial power in the 2025 timeframe" (Garretson 2010).

SPS over Japan

The only known country to commit to a schedule for producing energy from space with hard dollars (in this case, yen) committed is Japan. This happened in 2009 when JAXA (Japan Aerospace Exploration Agency) and the Japanese Ministry of Economy, Trade and Industry (METI) announced that Mitsubishi Electric Corp., IHI Corp. and associated companies would be awarded two trillion yen ($21 billion) in a project to build a 1 gw solar power generator in space within three decades for the purposes of beaming electricity to Japan (Sato and Okada 2009).

According to this plan, a research group representing 16 companies will spend 4 years developing the wireless power transmission station to be fitted with 4 km² of solar panels. One gw of energy collected from space was estimated to be equivalent to the production capabilities of a medium-size nuclear power plant, sufficient to power about 300,000 Tokyo homes. Its goal is to produce electricity at 8 yen p/kwh, six times cheaper than its current cost in Japan.

The strategy is to launch a smaller test satellite in 2015. The project's roadmap calls for a Japanese rocket to position a satellite into an LEO designed to test the beaming of energy from space through the ionosphere, the outermost layer of Earth's atmosphere. The next step, expected around 2020, would be to launch and test a larger, more flexible photovoltaic structure with 10 mw capacity, to be followed by a 25 mw prototype. The government said it hoped to have the solar station fully operational in the 2030s (Poupee 2009).

The Japanese National Space Plan

In June 2009, Japan had released its national space plan calling for a program "to lead the world in space solar power." For more than a decade, Japanese scientists had been investigating solar power satellites. In the 1980s, Hiroshi Matsumoto, radio scientist and Kyoto University president, was working with Kobe University's Nobuyuki Kaya launching rockets into the ionosphere to investigate what happens to microwaves as they travel through space. In May 2008, a team of researchers headed by Nobuyuki Kaya and NASA scientist John Mankins demonstrated power beaming over a distance of 148 km, between two Hawaiian Islands (Cyranoski 2009, p. 298).

One imaginative proposal is advanced by the research division of the Japanese construction giant Shimizu Corporation as perhaps the biggest public infrastructure installation ever constructed. This project would turn the Moon into a gigantic mirrorball manned by robots to provide energy to Earth. An array of solar cells would extend like a belt along the entire 11,000 km lunar equator. This belt would grow in width from a few to 400 km. The ambitious project would result in 13,000 tw of continuous solar energy being transmitted to receiving stations on Earth, either by laser or microwave.

Robots, remotely operated 24 h a day from Earth, would play a vital role in construction on the lunar surface. These machines and other equipment would be assembled in space and set down on the lunar surface for installation.

"A shift from economical use of limited resources to the unlimited use of clean energy is the ultimate dream of all mankind," the *Daily Mail* quoted the Shimizu Corporation's website. "Shimizu Corporation proposes the Luna Ring for the infinite coexistence of mankind and the Earth" (*Daily Mail* Reporter 2011).

In April 2011, a month after its tragic earthquake and tsunami, a top Japanese official was acknowledging that the country's space budget would take a hit as resources are diverted to recovery efforts, but felt confident that the government would be determined to maintain most space investment efforts. Hirofumi Katase, deputy secretary-general for the Cabinet Secretariat, Secretariat Headquarters for Space Policy, says, "The government is convinced that space utilization is something Japan cannot abandon. The long-term benefits are recognized" (de Selding 2011).

International Cooperation and Collaboration

Mark Albrecht was the executive director of the U. S. White House National Space Council from 1989 to 1992 and the principal advisor on space to President George H. W. Bush. In a *Space News* editorial, Albrecht writes:

> Realistically, absent another Cold War-like competition, the most promising scenario for successful human spaceflight beyond low Earth orbit will involve a truly international effort that relies on an integrated international and independent management construct.
>
> Unfortunately, there are no good examples upon which to model the U.S.-led International Space Station has left many partners bruised and suspicious of subordinating national treasure and pride to U. S. program management.
>
> The European Space Agency, which coordinates and aligns the activities of several independent national space activities, such as the French space agency CNES and the German Aerospace Center, is not much of a better template, as years of experience have shown parochialism and equity management to trump efficiency and accomplishment time and again.
>
> What is needed is an international space consortium where nations contribute critical pieces of technology and infrastructure based on demonstrated capabilities, most advantageous geographical location, level of resource commitment and embedded technology base, managed by an independent all-star team that is nation-blind, dedicated to success, free to make decisions without regard to nationality... and must be disbanded when the mission has been accomplished.
>
> This may seem Utopian and highly unrealistic... unless, of course, an asteroid is found to be on a certain collision path with Earth within 15 years, and all of this will seem simple and logical (Albrecht 2009, p. 19).

Concluding Thoughts

In writing his essay, Mark Albrecht was not thinking specifically of an international effort on behalf of space-based solar power, but the application is worth considering for the purpose of meaningful international collaboration.

The most pressing and immediate need for cooperation among spacefaring countries in the twenty-first century is in new energy production, and the most obvious destination for near-future space collaboration is in near-space. Collaboration on further exploration of the Moon and Mars and our entire Solar System will come in time, but we need the energy that will produce the capital to go there. Back home on Earth, things are not going well economically or environmentally.

The author concludes that one favor we can do for ourselves at this intermediate moment in space history is to strengthen our homeland economies. And we need to do something about the environmental devastation being caused as a result of the mining and burning of fossil fuels. We need shore up and secure our fresh water supplies and make sure that there is sufficient food to feed our growing population. Each of these basics is very much dependent on our people—all people—having access to clean and abundant energy.

Fortunately, we as a human race of aspiring spacefarers, don't have to go very far into space to secure Sun power in abundance.

References

Albrecht, M. 2009. Groundhog Day on Mars, *Space News International* 20(48): 19.

China Times Reporter. 2011. China unveils plan for solar power station in space. China Times, 2 September 2011. http://www.wantchinatimes.com/.

Cyranoski, D. 2009. Japan sets sights on solar power from space. *NatureNews*. http://www.nature.com/news/2009/091125/full/462398b.html. Accessed 12 May 2010.

Daily Mail Reporter. 2011. How the Japanese plan to turn the moon into a mirrorball: All of Earth's energy 'to be supplied by lunar ring of solar panels.' *Mail Online*. http://www.dailymail.co.uk/sciencetech/article-1390682/Plans-gigantic-lunar-ring-solar-panels-beam-energy-Earth-unveiled.html. Accessed 30 May 2011.

de Selding, P. B. 2011. Official: Japanese space commitment still strong. *Space News*. http://www.spacenews.com/civil/110413-japanese-space-commitment-strong.html. Accessed 30 May 2011.

Flournoy, D. 2009. Solar power satellites: Our next generation of satellites delivering the sun's energy to Earth. Address given before the International Conference on Space Information Technology (ICSIT09), Beijing, China, November 26, 2009, published along with the peer-reviewed papers of the conference by SPIE's *Optical Engineering*, SPIE 7651, 76513O, DOI:10.1117/12.855570.

Foust, J. 2010. Space solar power's Indian connection. *Space Review*. http://www.thespacereview.com/article/1721/1. Accessed 18 May 2011.

Gao, J., H. Xinbin, & W. Li. 2010. Solar power satellites research in China. *Online Journal of Space Communication*. http://spacejournal.ohio.edu/issue16/ji.html. Accessed 20 May 2011.

Garretson, P. A. 2010. Sky's no limit: Space-based solar power, the next major step in the Indo-U.S. strategic partnership. IDSA Occasional Paper No. 9. http://www.idsa.in/occasionalpapers/SkysNoLimit_pgarretson_2010. Accessed 30 May 2011.

Gopalaswami, R. 2010. Sustaining India's economic growth. *Online Journal of Space Communication*. http://spacejournal.ohio.edu/issue16/gopal.html. Accessed 20 May 2011.

Kalam, A. Harvesting space solar power through world knowledge platform. A teleconference address given at the China Energy Environment Summit, 27 August 2011. www.abdulkalam.com

Poupee, K. 2009. Japan eyes solar station in space. *AFP*. http://www.google.com/hostednews/afp/article/ALeqM5i8gMGQ65q2v3oVXxlLaYlckcUFdw. Accessed 30 May 2011.

Sato, S. & Y. Okada. 2009. Mitsubishi, IHI to join $21 billion space solar project. *Bloomberg*. http://www.bloomberg.com. Accessed 15 September 2009.

Shimbun, Yomiuri. "Space-based solar power set for 1st test," Yomiuri Shimbun, 23 January 2011, http://www.yomiuri.co.jp/.

Chapter 9
What Is Worrisome About Solar Power Satellites?

Abstract This chapter addresses some technological constraints, implementation costs and other challenges facing space solar power satellite systems.

Launch to Space

Hearing space scientists say, "The science behind space-based solar power is sound," and "Solar power satellites are technically feasible," must not be interpreted to mean, "The hard work is all done." As energy-generating satellites are being positioned for launch, informed professionals readily acknowledge the multiple technical and non-technical issues yet to be resolved, many of which can ultimately be addressed only in practice.

The biggest obstacle to space-based solar power is the difficulty and expense of putting satellites into orbit using today's technology and business models. The lack of a regularized transportation system is widely thought to be the single most significant factor holding back near-term implementation of space solar power. The current high cost is attributed to the small number of satellites destined for space each year, the limited number of launch vehicles available to do this work and the fact that almost none of the launch vehicles is reusable, which underscores the reality that space is not yet taken seriously as a commercial destination.

China for the first time in 2010 surpassed the United States in number of launches when it transported only 15 satellites. Taken together, our spacefaring nations launch only about 100–120 satellites of any type each year, and these are launched by a handful of countries—principally the United States, China, European Union, Russia, Japan, India and Israel.

No matter how impressive the designs for space-based solar power systems, these concepts and business plans will go nowhere until there is economical, reliable and frequent access to space. Lowering the cost to orbit is expected to prompt entirely new commercial enterprises, some of which will be transformative for the countries and businesses that pursue them. Proponents of energy from space argue that the new solar power satellite market alone will be big enough to bring down launch costs.

Assembly in Space

A second concern relates to assembling, managing and maintaining solar power satellite operations in space orbit. Like the ocean depths, outer space is not conducive to human survival; thus, once rockets lift the basic material components into orbit, robots will most likely be the hands-on managers of these operations, extending human sight, reach and intelligence by means of electromagnetic communications and control.

The Mars exploration rovers Spirit and Opportunity—examples of this kind of interface—have been followed on TV as if they were NASA rock stars. Launched in 2003 and landing on Mars in 2004, these two robotic extensions of humankind have been searching for answers about the history of water on this distant planet some 60 million km from Earth (NASA 2011).

Equipped with wheels, these rovers were expected to spend their life traveling over a distance of 1 km in different parts of the Martian terrain, performing on-site geological investigations. Each carried a panoramic camera for taking pictures of the local terrain and a spectrometer for close-ups of rocks and soils. A robotic arm, capable of moving like a human arm with an elbow and wrist, could place instruments directly up against rock and soil targets of interest. Eventually, one rover became stuck and was turned off in 2011, but both traveled farther and remained responsive to human command far longer than expected. And the full-color images relayed to Earth on a daily basis were followed with interest by the media and audiences everywhere.

The Mars rover experience lends credence to the idea that space structures of considerable mass and complexity can be remotely assembled, monitored and managed by human controllers safely on the ground.

Wireless Transfer of Energy

Possible negative health and environmental effects of beaming energy from space satellites to Earth antennas are a matter of public concern. Anticipating and preventing such effects, if any, is a necessary priority for the emerging space solar power industry.

A typical reference design, circa 2003, involves a satellite in geosynchronous orbit, with photovoltaic arrays of several kilometers continuously pointed toward the Sun. Energy collected is converted into radio frequencies of 2.45 or 5.8 GHz, which is considered most suitable for transmission through Earth's atmosphere. Targeting a pilot signal on Earth, these frequencies are beamed via a wireless power transmitter to a designated antenna—also several kilometers in size—on the ground. The rectifying antenna converts the energy into electricity compatible for distribution on the terrestrial grid. Such an installation can deliver as much as 5–10 gw of electrical power. At the location where the beam encounters the ground, intensities are expected to be about 1/16 of noon sunlight (SBSP Study Group 2007, pp. 7–8).

In 2007, a space-based solar power study group was commissioned by the National Security Space Office of the U. S. government to update NASA's 1997 "Fresh Look" study on energy from space. The office was concerned about potential political conflicts (wars) arising as a result of increasing global population and declining energy resources. In addition to energy security, the group was asked to consider environmental, economic, intellectual and space security as well. The group found that "when people are first introduced to this subject, the key expressed concerns are centered on safety, possible weaponization of the beam and vulnerability of the satellite, all of which must be addressed with education."

> Because the microwave beams are constant and conversion efficiencies high, they can be beamed at densities substantially lower than that of sunlight and still deliver more energy per area of land usage than terrestrial solar energy. The peak density of the beam is likely to be significantly less than noon sunlight, and at the edge of the rectenna equivalent to the leakage allowed and accepted by hundreds of millions in their microwave ovens. This low-energy density and choice of wavelength also means that biological effects are likely extremely small, comparable to the heating one might feel if sitting some distance from a campfire (SBSP Study Group 2007, p. 26).

Land Use

Ground positions are becoming scarce for new wind and solar farms, as well as for highways, gas pipelines, airports, hospitals and prisons. Thus, the question arises, "Where can one find 5 square miles of protected Earth on which to put a space solar power rectenna?"

Locating a good site may not be as challenging as one would first assume. A unique characteristic of microwave-transmitted energy from space is that agriculture can coincide with the rectenna site to the advantage of both. A rectenna can be erected directly above a terrestrial solar farm, doubling and perhaps tripling its capacity, or above a coal or gas-fired power plant to reduce its dependency on fossil fuels.

How is this possible? Microwave receiving rectennas are designed to absorb almost all of the beamed energy but allow the larger percentage of ambient light to pass through. In rejecting excess heat, these space solar antennas can retain sufficient warmth and light to power large greenhouse complexes for year-round flower and vegetable production, to support an ongoing feedlot operation for raising livestock or to sustain the atmosphere needed to keep farm ponds active 12 months a year for the production of fish or algae.

At the International Space Development Conference in 2011, Ohio University students presented a digital visualization of a space solar power application that converted 5 square miles of abandoned strip mine land in southeastern Ohio into a working rectenna. According to the technical brief and business plan associated with the design, this site would be sufficient to supplement and eventually replace the productive capability of a coal-fired plant owned by American Electric Power rated at 1 gw.

Another of the Ohio University student designs employed abandoned oil well drilling platforms off the coast of southern California for use as rectennas for powering saltwater desalination units pumping fresh water to shore. This same system was also sized to transmit wireless electric power sufficient to power a city of 45,000.[1]

Satellite Collisions

Some observers, fearing that space is getting crowded, worry that satellites will start crashing into one another. Although these incidents are rare, they do happen.

The Space Data Center, established on the Isle of Man by the Space Data Association in 2009, maintains an automated space situational awareness facility that works to reduce the chance of satellite collisions and frequency interference on a global basis. The need for such a facility was prompted in part by events that occurred in space in 2007 and 2009. The first was a Chinese military demonstration of a "kinetic kill vehicle" that destroyed one of its own retired weather satellites, the Fengyun-1C. The impact exploded the satellite into 3,000 separate pieces of debris 10 cm or larger. The second occurred when an orbiting Iridium 33 (satellite telephone) spacecraft collided with a defunct and wandering Soviet-era Cosmos 2251 satellite. The collision created 2,100 additional pieces, all of which are moving at high speeds in orbital space (Schrtz 2010, pp. 172–180).

Prevention requires two different types of preparedness. The first is to assure that no satellite crash is intentional, i.e., a human-directed event. In such a case, the recommended solution is nation-to-nation or corporation-to-corporation diplomacy. For accidental collisions, monitoring and managing the space environment and data sharing are the recommended preventive measures. In both cases, liability conventions should be in place (or at least there should be a process for adjudication of differences).

It may be helpful to explain why satellite collisions rarely happen: one of the most crowded satellite orbits is the geosynchronous arc located 36,000 km above

---

[1] To view these and other space solar power designs see http://spacejournal.ohio.edu/issue17/main. html. Other visualizations are being solicited in the 2011–2014 SunSat Design Competition being sponsored by the Society of Satellite Professionals International, the National Space Society and the *Online Journal of Space Communication* hosted at Ohio University. Click on http://sunsat. gridlab.ohio.edu/.

Earth's equator. Within the 360° orbit, the ITU makes no satellite assignment that is less than one degree apart, which means that spacecraft placed in GEO will be no closer than 773 km, the approximate distance between east and west coasts of Panama. With such a separation, satellites orbiting with Earth each 24 hours at comparable approximate speeds of 11,000 kph are not likely to bump into one another. Satellite spacing within other orbits receives similar monitoring.

Space Debris

A much bigger concern are the 5,100 pieces of debris caused by the Chinese A-Sat and Cosmos/Iridium events (and others released by humans and by nature) that are still orbiting in a place where numerous Earth observation, meteorological and other satellites are located.

According to Nicholas Johnson, NASA's chief scientist for orbital debris, the Department of Defense is responsible for tracking materials larger than 10 cm, while NASA is responsible for anything smaller. By using ground-based telescopes and radar, NASA is tracking as many as 300,000 particles from the two collisions (and others), yet has no capability to remove them.

Interviewed by *Space News*, Johnson said, "I've been the U. S. technical expert on orbital debris at the United Nations for the last 14 years.... We have guidelines. If everybody follows the guidelines, the change in the environment will be very, very modest. One of the good things about debris is that by and large debris has characteristics that allow it to come back to Earth more quickly than satellites" (Werner 2010, p. 18).

Solar Storms and Flares

In April 2011, a commercial satellite fleet operator in Thailand announced that its Thaicom 5 satellite had suffered a 4-hour service outage due to an apparent electrostatic discharge. The satellite, located in GEO at 78.5° east longitude, had automatically placed itself into safe mode and pointed itself toward the Sun to maintain electrical power. A company spokesman explained, "[A]n electrostatic discharge cannot be predicted in advance, and its occurrence is quite rare" (de Selding 2011b, p. 3). The satellite was not damaged.

An earlier in-orbit failure of the Galaxy 15 satellite owned by Intelsat was initially blamed on an electrostatic discharge. In that case, the satellite was unable to respond to commands and began a 6-month uncontrolled drift along the geostationary arc before it was brought back under ground control. Intelsat reported in January 2011, following a complete checkup, that Galaxy 15 appeared to be in good health. *Aviation Week & Space Technology* noted, "Of the 120 potential root causes identified, only two remain. Solar flares, the long-rumored culprit, are not one of them" (Taverna & Morring 2011, p. 38).

In brief, solar flares trigger the ejection of coronal mass (an eruption of gas) from the face of the Sun that is propelled through space. Our Sun follows an approximate 12-year cycle of such activity but is thought to now be entering a period of relative quiet (perhaps the least activity in 80 years). These flares can disturb Earth's ionosphere, the uppermost part of the atmosphere, and have led to radio blackouts. Such flares can also create static electricity that can discharge and short circuit vital electrical components on spacecraft.

Solar storms can be responsible for the magnetic fields and charged particles that have washed over power lines on the ground, melting transformers and cutting power to those connected to electrical utility grids (Clark 2009, pp. 27–31). These, too, are rare events that cannot be easily predicted and even less easily protected against.

Signal Interference

Communication providers are protective of the electromagnetic frequencies assigned to them and watchful of any applications or events that might compromise their signals. A 2011 example is the case of hybrid satellite/terrestrial broadband wireless provider Light Squared, a telecommunications company that ran into a storm of protest when the L-band (2 GHz) radio spectrum it proposed to use was thought to cause interference in an adjacent spectrum assigned to global positioning systems. When both government and industry users of GPS aired concerns, the company was sent scrambling for an alternative frequency plan (de Selding 2011a, p. 5).

Communications satellites transmit electromagnetic waves that convey voice, data, video, navigation and timing signals. Solar satellites transmit electromagnetic beams that convey energy and electrical power. Research has not determined whether communications signals and energy beams are compatible. The author put the question to Dr. Paul Werbos, IEEE Fellow and member of the National Science Foundation Energy, Power and Adaptive Systems (EPAS) group and a number of other informed professionals. A search of IEEE's *Journal on Microwave Theory and Techniques* led to an inconclusive answer.

The compatibility (or incompatibility) of energy and communication is an example of an unanswered question that could make a big economic and political difference for solar power satellites. If space power and space communications services can originate from the same space platform, the comsat industry will more readily embrace energy as its next new market. If not, launching an energy-from-space program will be a greater challenge because of all the interference issues that could arise.

Concluding Thoughts

Whether it relates to communications and media, imaging and remote sensing or geo-positioning and location services, launching a new space business is never going to be easy. As with launch of any global enterprise on Earth or space, there will always be causes for worry. The energy-from-space initiative will have to find its place and address its challenges as it moves forward. Since its product is greatly needed, this development will have lots of encouragement and help along the way.

References

de Selding. P. B. 2011a. LightSquared suffers setbacks on two fronts. *Space News*. http://www.spacenews.com/satellite_telecom/110617-lightsquared-setbacks-two-fronts.html. Accessed 20 June 2011.

de Selding. P. B. 2011b. Thaicom 5 quickly restored to service. *Space News*. http://www.spacenews.com/satellite_telecom/110422-thaicom5-service-restored.html. Accessed 22 April 2011.

Clark, S. 2009. Cosmic Katrina. *BBC Knowledge*. September/October.

NASA Jet Propulsion Laboratory, California Institute of Technology. 2011. *Mars Exploration Rover Mission*. http://marsrover.nasa.gov. Accessed 30 June 2011.

SBSP Study Group. 2007. Space-based solar power as an opportunity for strategic security, phase 0 architecture feasibility study: report to the director, National Security Space Office interim assessment. http://www.nss.org/settlement/ssp/library/nsso.htm. Accessed 15 June 2011.

Schrtz, E. I. 2010. A brief history of space junk. *Wired*. June.

Taverna, M. A. & F. Morring, Jr. 2011. Prodigal son returns. *Aviation Week & Space Technology*. January 24/31.

Werner, D. 2010. Profile: Nicholas Johnson. *Space News*. www.spacenews.com. Accessed 10 May 2010.

Chapter 10
Top Ten Things to Know About Space Solar Power

Abstract This is the frequently asked questions (FAQ) chapter, the place you can go for answers to the top ten questions asked about space-based solar power—questions such as, "Why are solar power satellites needed, are they feasible and when will we see them?"

#1: Isn't This Just Science Fiction?

Answer: Once upon a time, solar power from space was the subject of science fiction; however, every year this idea gets closer to science fact. Science fiction writer Arthur C. Clarke is often quoted as saying, "Any sufficiently advanced technology is indistinguishable from magic."

Clarke (1917–2008) lived long enough to see a good deal of magic become science reality. In 1945, long before space satellites were ever thought practical, he wrote about a place in space some 36,000 km above Earth where an artificial satellite would orbit at the same speed as Earth's rotation. That is, the satellite would appear to be stationary with respect to any point on Earth's surface. That place, now commonly called the Clarke Belt, is where the majority of communications satellites are located that provide essential voice, video, data and other services to Earth.

In his *2001: A Space Odyssey*, Clarke and his motion picture collaborator Stanley Kubrick depicted a commercial space plane delivering passengers to a huge, wheel-shaped space station featuring hotels, restaurants and videophone booths onboard. This fiction was not realized within the exact timeframe depicted, but by 2011, space planes are indeed flying passengers into near-space, and astronauts have inhabited the International Space Station in low Earth orbit for more than a decade (Fig. 10.1).

Another science fiction author, Isaac Asimov, played with the idea of solar power collected in space and beamed to Earth in his 1941 short story, "Reason." Dr. Peter Glaser of Arthur D. Little, Inc. developed a detailed proposal for such a project in 1968 and was issued a patent on the concept in 1973. Indeed, in 2011, the basic idea for using large satellites in geosynchronous and other orbits to collect gigawatts of energy that will arrive just in time to rescue the planet from our dependence on fossil fuels still has a science fiction aura about it. But as this *Solar Power Satellites*

D.M. Flournoy, *Solar Power Satellites*, SpringerBriefs in Space Development,
DOI 10.1007/978-1-4614-2000-2_10, © Don M. Flournoy 2012

Fig. 10.1 The visualization pictured won the grand prize in the NASA-NSS Student Space Settlement Design Contest at the 2011 International Space Development Conference, Huntsville, Alabama. The Hyperion Space Settlement, with a diameter of 1.8 km, is powered by space solar energy. The winners were high school students from Punjab, India (National Space Society 2011)

book illustrates, the vision behind the idea of solar power satellites is now much closer to the point of demonstrated reality.

#2: Why Is This Important to Me?

Answer: As citizens of Earth, we have allowed ourselves to become dependent on electricity for almost everything we do, even though we are fully aware that some 90% of all electrical power is produced from a rapidly depleting fossil fuel resource stock, the burning of which is creating havoc with our environment and our climate. This situation is not unlike that described by Sony Corporation of America CEO Howard Stringer who (in jest) said of investor expectations just prior to the bursting of the year 2000.com bubble, "We all knew it was coming but we wanted to get in and get out before it happened."

We all know that terrestrial power grids are fragile and can go down at any minute for any length of time. The rule of thumb for the financial services industry, which spends a lot of money on disaster recovery solutions and insurance, is that banks whose telecommunications lines are down for a week will be out of business. It is not hard to imagine the result of a natural or manmade disaster/switching failure causing multiple interconnected electrical grids to fail. When that happens, the most basic functions of modern society will no longer be available.

Think of this: government offices, businesses, hospitals, schools and residences will go dark. Airports, train stations and bus depots will no longer function; gasoline stations will no longer pump gas. Computer networks, ATM machines, heating and cooling systems and refrigeration will no longer switch on. Without electrical power, radio and TV will not work, nor will e-mail, cell phones or game players. All the social media apps will instantly vaporize. When the lights go out, civilization as we know it will cease to exist.

Proponents of space solar power sometimes describe what they are doing as "mining up" or "drilling up" for energy. These are unfortunate analogies from an industrial age when it was still acceptable for wealth to be created at the expense of sustainable life on planet Earth. Reaching up and harvesting the energy that is always there just outside our atmosphere is to tap into a direct, clean and continuing baseload electrical power supply. When viewed in terms of the lifespan of human civilization, our Sun is in no way diminished by the capturing of its rays. Sun power is a virtually inexhaustible resource.

A more appropriate analogy is to describe our Sun as a fully-charged storage battery whose capacity is massive. The Sun's energy is a source sufficient to reduce, if not entirely replace, our dependence on fossil fuels. Thus, if we are smart and careful, and do not foul the space above us, our civilization is sustainable from this source for the foreseeable future.

#3: Who Is Taking the Initiative with Solar Power Satellites?

Answer: Were an asteroid proven to be on a collision course with Earth—and there was something humankind could conceivably do to avert that disaster—chances are very good that all nations would come together to work for the planet's defense. Given that few nations are currently acknowledging that our Earth has a serious energy-access/energy-use problem, and even fewer feel they are in a political or economic position to do anything that might make a difference, the disposition toward there being international cooperation in drawing down power from space is not yet evident.

To make effective use of its growing mastery of space, China already has a plan to complete its first 100 kW solar power satellite in low Earth orbit as a demonstration project in 2025. It expects to have 100 MW electricity generating capacity in place by 2035, with the first commercial-level SPS system in operation in geosynchronous orbit by 2050.

In the meantime, to meet its increasing demand for energy, China reports that it will be relying on coal. According to China's 2009 National Bureau of Statistics Report on Energy Development, the country consumed 2.85 billion tons of coal in 2008. Its total energy consumption of standard coal is expected to increase to 3.5 billion tons in 2020, to 4.2 billion tons in 2030 and to 5.0 billion tons in 2050. Some 85% of its growth in demand will come from fossil fuels, from nuclear power and from hydropower. In 2050, the remaining 15% of its energy will come from wind power, bio-energy, terrestrial solar power, tidal energy and space solar power (Ji et al. 2010, p. 1).

Japan is among the globe's long-term energy planners. For decades, Japanese scientists and engineers have been considering energy from space as a potential solution for a small country with high energy needs but few natural resources. Although Japan invested heavily in nuclear power, research continued on the clean energy option from space. In the year before the Fukushima nuclear disaster, Japan made news by being the first nation in the world to publically commit substantial financial resources (two trillion yen) to developing a practical solar power satellite solution (Sato and Okada 2009).

The Japan Aerospace Exploration Agency (JAXA) plans to launch into geostationary orbit a solar power generator that will transmit 1 GW of energy to Earth, equivalent to the output of a large nuclear power plant. At the time of the announcement, the government's goal was to launch a smaller satellite fitted with solar panels in 2015 to test the beaming of electricity through the ionosphere, the outermost layer of Earth's atmosphere. It expected to have the station operational in the 2030s, delivering electricity to the island nation at a competitive cost of $0.09 p/kWh.

Sources predict that India will need some 800 GW of additional electricity per year going forward, compared to the 600 GW predicted for China. (Kollipara 2009) As a spacefaring nation, India has had designers at work for more than two decades on affordable space solar power systems, in part because the government has figured out that all the known non-carbon energy sources combined—including hydropower, wind power, Earth solar, geothermal and biomass—would be insufficient to meet even its most modest growth expectations.

Indian leaders do not view their country as among those remaining forever in the "developing world" category. In 2010, there was a government constraint placed on use of coal, yet the targeted GDP growth rate was set at 7%. According to Raghavan Gopalaswami, a senior government advisor, this metric "would call for power capacity growth at 5.42% every year from 2010 onwards on a base of 160.7 GW. By 2052, the installed power capacity should be about 1,476 GW for sustained 7% GDP growth." How then will this power capacity shortfall of 554 GW be filled? His conclusion is that "a non-constrained, non-terrestrial source of energy, namely Space Solar Power," will be its path forward (Gopalaswami 2010).

As the momentum builds for solar power installations in space, whether through government or private enterprise, it can be predicted that Canada, the European Union, Russia and other spacefaring nations will wish to contribute their experience and expertise and perhaps their financial resources. But, for the moment, none of these countries apparently sees exhaustion of terrestrial energy sources and destruction of the human habitat as an "imminent threat."

#4: Why Not Just Focus on Terrestrial Energy?

Answer: This is a commonly asked question, and a good one. Non-polluting renewable energy sources are indeed available on Earth. And we will need them all.

As of 2009, the power consumed worldwide at any given moment was about 12.5 trillion watts (tW), according to the U. S. Energy Information Administration, as quoted by Jacobson and Delucchi in *Scientific American*. This agency projected that in 2030 the world will require 16.9 tW of power generation as global population and living standards rise. The United States alone will require about 2.8 tW (Jacobson and Delucchi 2009, p. 60).

With economic incentives and environmental and regulatory pressures, more of the cleaner energy alternatives are appearing in the marketplace. In addition to the old hydroelectric standbys, plausible near-term supplements to coal, gas, oil and nuclear energy production now include terrestrial solar, wind and biomass. Further out we will see geothermal, fuel cell, tidal power and other clean energy solutions being added to the mix.

As for terrestrial solar, only about half of the energy emitted by the Sun actually reaches Earth's surface. Some is absorbed and scattered about, but a considerable portion is reflected back into outer space. Even so, scientists have demonstrated that the amount of the Sun's energy present is far greater than the energy currently produced in the burning of fossil fuels. Thus, it would appear that such renewable solutions as solar, wind and biomass would be sufficient. This assumption is incorrect, largely because energy direct from the Sun is not always available when and where it is needed, and it is not in a readily usable form.

The Sun's rays on Earth are intermittent, relatively weak and diffuse. Only about 20% of that potential energy is accessible in daylight hours in the best of locations, which means that, for on-demand service, energy production units must be connected to a power distribution system. On-site battery or other storage systems are required for a continuous flow of electricity. Multiple solutions to these challenges are of course well underway in the form of large-scale terrestrial solar production systems in suitable sites around the world. These are greatly needed. In aggregate, however, they are predicted to fall far short of the goal to replace our dependence on the extractive industries.

Wind turbines are promising sources of alternative energy as well, since they have near-zero emissions of greenhouse gases and air pollutants over their entire life cycle, including construction, operation and decommissioning. Jacobson and Delucchi note that nuclear power results in up to 25 times more carbon emissions than wind energy when reactor construction and uranium mining, refining and transportation are factored into the equation (Jacobson and Delucchi 2009, p. 59).

Biofuels will also be part of the eventual solution. Ethanol made from corn is used as a petroleum substitute in automobiles. Biodiesel from sunflower seeds has proven to be a suitable power source for farm tractors. A Gulfstream jet in 2011 flew from New Jersey to France using fuel produced from the camelina plant seed. Yet, when taking into consideration the intensive farming, harvesting and processing involved in the production of fuel from biomass, its ultimate contribution to CO_2 reduction is thought to be minimal and may indeed make matters worse. This energy is unlikely to be cheaper, and one of its side effects may be to increase food costs due to overuse of agricultural land for the production of fuel.

For the foreseeable future, all countries and communities will have little choice but to draw on the energy resources they have, polluting or not. It does not seem realistic to think that demand will decrease, or that provider efficiency and user conservation will make a definitive difference. These are some of the reasons why a solar power satellite solution remains attractive to the few who are aware of it. (For an authoritative source on this topic see John Strickland's article "Space Solar vs. Base Load Ground Solar and Wind Power" in Issue No.16, Online Journal of Space Communication, Winter 2010 www.spacejournal.org).

#5: Will These Sunsats Be Like Comsats?

Answer: Had the United States and other spacefaring nations the foresight in the 1960s to launch solar power satellites, as well as communication satellites, universal access to energy in every part of the world today could very well be as easy and inexpensive as accessing voice, video and data.

Had nations joined together for the common cause of energy production and distribution, as they did in 1963 with the creation of the non-profit international satellite collaborative (Intelsat) for global communications in which no nation was to be excluded, space solar power could very well be seen today as a "human right," and there would be a means to achieve that goal.

The ITU, headquartered in Geneva, Switzerland, is an organization within the United Nations where governments and the private sector coordinate telecom networks and services. At its World Summit in 2003, the ITU issued a "Declaration of Principles" for building the Information Society; that is, "to build a people-centered, inclusive and development-oriented Information Society, where everyone can create, access, utilize and share information and knowledge, enabling individuals, communities and peoples to achieve their full potential in promoting their sustainable development and improving their quality of life…" (World Summit 2003, p. 1).

It is conceivable that, within the United Nations, a new "International Energy Union" could be established, with purposes and principles similar to those of the ITU, coordinating energy networks and services, and facilitating energy generation and distribution as a global enterprise.

Access to energy has emerged as a matter of social equity, since energy is key to all economic and social development. In its earliest form, telecommunications was the exclusive domain of governments and the military. But today, it is a globally competitive multi-billion-dollar industry serving governments, military and all society. Now, Intelsat is operating a commercial satellite fleet capable of providing advanced communications to almost anywhere in the world. Energy is not communications, of course, but many of the infrastructural requirements in space and on Earth are similar, and many of the safety and environmental concerns are the same.

It is also possible that a new type of international solar satellite collaborative will be established, drawing on Intelsat's 40 years of experience, targeted to the special needs of nations for access to clean, economical and abundant energy over the long term.

#6: How Will Solar Power Satellites Get into Space?

Answer: In 1977, *Voyager 1* and *2* were launched from Cape Canaveral just months apart aboard Titan 111E/Centaur rockets. These unmanned NASA space probes were on their way to tour and transmit photographs from the planets of our Solar System.

The two *Voyagers*' journey to the planets was to have taken 10 years. Thirty-four years later, even though they have entered the heliosphere—the wide span of space that marks the edge of our Solar System—the two spacecraft are still operational. Their cameras have been turned off, but NASA is still in communication with them. At 15 billion kilometers from Earth, they are the farthest-traveled human-built objects in space, so distant from the Jet Propulsion Laboratory in Pasadena, California, that it takes 13 hours for an electronic message traveling at the speed of light to reach them, and another 13 hours to get a message back (Klotz 2010).

The *Voyager* craft were launched into space by a Titan III rocket powered by the combustion of onboard solid or liquid fuel. The family of U.S. Titan expendable rockets carried satellites and other cargo into space some 368 times between 1959 and 2005 (Poynter & Lane 1984). The Titan IVs were capable of placing 18,000 kg of mass into LEO and 4,500 kg into GTO.

The crucial stage of space launch is liftoff, when the launch vehicle will use most of its fuel in the first few minutes to escape Earth's gravitational pull. This is the principal reason why the majority of a launcher's mass is the fuel it carries, not the payload itself. Using *Voyager* as an example, the satellite payload likely represented only 1–2% of Titan's weight. Satellites are designed to be small and light. Carefully tucked into the launch vehicle, they will deploy their antennas and solar arrays only after they have been successfully lifted into orbital space.

Once they are in the weightlessness of space, solar cells will normally be used as the principal power source, since these onboard arrays can collect and convert sunlight directly into electrical energy. For the *Voyagers* traveling to the far reaches of our Solar System, NASA engineers knew that the solar cell solution would not work. Jupiter is so far from the Sun that it gets only 1/25 as much sunlight as Earth, and Saturn is even farther still. Instead, these craft carried onboard a small, low-weight nuclear power source called a radioisotope-thermoelectric generator, by which plutonium reacting with oxygen creates plutonium oxide that serves as their source of propulsion (Poynter & Lane 1984, pp. 6–7).

Looking forward, it seems obvious that enormous amounts of terrestrial energy will continue to be required to launch the components that will comprise the infrastructure for the next-generation solar power grid. In-space energy will be required to transport these materials from low Earth orbits into higher orbits. The 1979 reference architecture, for which the eminent space scientist John Mankins held some responsibility, envisioned the deployment of as many as 60 solar power satellites in geostationary orbit, producing from 5 to 10 GW of continuous energy (Belvin 2009). Using direct-drive high-performance solar arrays, solar electric propulsion systems will meet many future needs for in-space transportation at considerably less cost (Howell 2005).

Old energy will continue to be consumed in establishing and maintaining a new and sustainable solar power energy resource in outer space. The paradigm, however, will be quite different, as the new energy produced will be clean and the source infinitely renewable.

#7: Is a Solar Power Satellite Grid Economically Feasible?

Answer: A public school teacher—on behalf of her eighth-grade student researching a paper on space solar power—asked, "How much would SBSP (space-based solar power) cost?" Three long-time National Space Society members, John Strickland, Darel Preble and Paul Werbos, responded to the question and shared their thoughts with the author. Strickland writes:

> A city with a power demand of 1 GW needs 24 GWh, just as your house might need 24 kWh if it only used 1 kW of power on average. A gigawatt is a million kilowatts. The Sun does not shine at night, so up to 21 GWh of power would have to be stored for a single average winter night. Typically a baseload kilowatt-hour of power costs about 5–12 cents. A kilowatt-hour is used when ten 100-W incandescent light bulbs are lit for 1 h.
>
> Space solar is a perfect source of baseload electricity since the Sun shines all the time in space and does not need any storage and does not require any fuel. Current estimates to provide baseload energy using solar and wind are as much as $60 billion a gigawatt, due to the huge energy storage requirements. Other standard sources cost about $2 to $5 billion a gigawatt. With reusable rockets, costs would be comparable to other baseload generating systems, surely below 15 cents per kilowatt" (Strickland 2011).

Preble writes that the key to understanding the question is to realize that construction and launch costs are currently too high. He explains that only a larger launch market can lower these costs, emphasizing that "space solar power is the only market we know of with the volume required to enable the low-cost space transportation system necessary for a successful business case. It is sort of a chicken and egg situation—you have to have one to get the other. Low cost transportation to GSO is critical; launch costs to GSO must be reduced from thousands of dollars per pound to about $150/lb. If the market demand were there, the price would then be about $50/lb to LEO or about $100/lb to GSO; and that is with very near-term technology. This theme repeats through other costs" (Preble 2011).

Werbos cautions, "We cannot be sure of what it *really* costs until we build it. Until then, we are estimating. The cost will be very different depending on the design. Each design is made up of different 'pieces' (Really subsystems and tasks). The cost depends on what we assume about the costs of the pieces.

> There has been no really thorough, credible life-cycle cost estimation for any design for energy from space since 2002 or 2003, when the last government funding was available for that purpose, subject to high-level review and oversight. That work from SAIC (Science Applications International Corporation) reported costs of 17 cents/kWh for the *lowest cost* variation of space solar power that they studied. More important, they said that this is a correct statement only after future improvements in the technology for subsystems and tasks—ambitious but attainable improvements, such as $200/lb-LEO and $200/lb LEO-to-GEO. We have reason to believe that new designs, beyond those studied in 2002, could get to lower costs—probably to 10, 5 or even 3 cents, *but we need that low-cost access to LEO,* and that is now a life and death crisis for the entire future of humans in space (Werbos 2011).

In 2009, *Scientific American* published "A Path to Sustainable Energy," an article in which the authors noted that the average cost of conventional power generation in the United States was about 7 cents/kWh and was projected to be 8 cents/kWh in 2020. "Power from wind turbines, for example, already costs about the same or less than it does from a new coal or natural gas plant, and in the future wind power is expected to be the least costly of all options.… Overall construction costs for a WWS (wind, water and solar technology) system might be on the order of $100 trillion worldwide over 20 years, not including transmission" (Jacobson and Delucchi 2009, p. 64).

In 2011, the capital cost estimates for coal and nuclear power plants by the Energy Information Administration were 25–37% higher than in 2010. These increases reflected the rising costs of capital-intensive technology in the power sector, higher global commodity prices and the fact that there are relatively few construction firms with the ability to complete such complex engineering projects as a new nuclear or advanced coal power plant (U. S. Energy 2010a).

Some proponents believe that space solar power systems may already be competitive with nuclear power when one takes into consideration that a single 1 GW nuclear power plant will cost some $4 to $10 billion to construct over a period of 5–10 years. Processed uranium as fuel is increasingly scarce and expensive and so is the cost of operating these plants safely. In 2008, Britain decided to build up to ten additional nuclear power plants by 2020. These were each estimated to cost $4.5 billion, not including disposal of their toxic waste (Porter 2008).

> At that time, Britain had not yet found a way to dispose of its nuclear waste. The Government was looking for a way to remove the accumulating waste being locally stored on power plant sites to a deep geological repository, but the time frame for that happening was thought to be 25 years. Also not included in the projections were the unaccounted decommissioning costs, estimated at billions of dollars, that future taxpayers will face as each of these plants reaches the end of its operational lifetime.

In a blog on European energy policy, Andrew McKillop wrote in 2011, "While we do not know and will not know the real cost of the Fukushima 4-reactor meltdown, because a period of 10 years is about the minimum needed to get a handle on it, the economic damage and loss from the Chernobyl 1-reactor meltdown has been relatively well costed—over the years since it happened in 1986. At a minimum and in today's depreciating and devaluing dollars, the cost ballpark starts at about $250 bn" (McKillop 2011).

#8: How Is Solar Energy Transmitted to Earth, and Is It Safe?

Answer: Solar power satellites will collect the solar photons found in abundance in space and beam those photons to Earth as an electromagnetic wave. These are the same types of waves the comsat industry has successfully and safely employed to deliver video, data and voice communication to and from Earth for 40 years. In this case, though, the primary purpose of the satellite is to deliver energy, which in high volume and in concentrated forms must be managed with care.

In orbit, the Sun's energy is collected by arrays of photovoltaic (PV) cells similar to but much larger than those used to power the downlink signals of communications satellites. These space antennas, consisting of semiconductor "solar cells" for converting sunlight photons into voltage, will be several kilometers in diameter. Concentrating photovoltaic (CPV) systems can also be used to increase the efficiency and amount of light converted into electricity. Heat that develops at these collection points is radiated back into space, an advantage over terrestrial PV systems.

Onboard the satellite is a wireless power transmitter that will beam the energy to Earth in a manner quite similar to that used in RF (radio frequency) communications. A common microwave frequency in current use for digital audio/radio transmission to automobiles on Earth is at 3 GHz, and a frequency used for video transmission to cable companies is at 4 GHz. The two microwave frequencies most commonly considered suitable for power transmission are at 2.4 and 5.8 GHz. Decades of using these frequencies in space and on Earth have raised little or no concern about environmental and health issues.

On Earth, the frequency beams of space solar power will be collected by a rectenna, used to convert the energy conveyed in the electromagnetic wave into electricity. For normal applications, this electrical power will be fed directly into a commercial power utility interface. The sizing of antennas will be in proportion to the transmitting satellite's distance from Earth, power levels and electromagnetic frequencies used. It can be expected that these rectennas will require a designated receiving area of 1 km or more.

The method for transmitting electricity over the long distances from space was invented by Peter Glaser, who was granted a patent for this process in 1968. This was not a new idea, however, since wireless power transmission was undergoing laboratory testing in the early 1900s by Nicola Tesla, inventor of alternating electrical current. AC today serves as the principal means by which electricity is distributed via wires, though scientist and inventor Thomas Edison at the time favored direct current (DC), a different approach to distribution.

In answer to questions of safety, the principal concern about microwave transmission is one of heat. Just as people, wildlife and plant life can be negatively affected by over-exposure in the noonday Sun, so can they be affected if they are unprotected in the footprint of a concentrated beam of solar energy. The two most commonly recommended solutions are to fence off the area, as is universally done in the case of coal-fired and nuclear plants, and to widen the beam so that the heat of the Sun will never be greater than a midday visit to the beach.

In the United States, the Occupational Safety and Health Act has set standards for such exposure. In placing and maintaining Earth antennas, such requirements are to be anticipated and followed. One common technological solution for "failsafe beaming"—assuring that the transmission does not stray from its designated spot on Earth—is to establish from the center of each rectenna plot a continuous pilot beam that tracks the solar power satellite in its orbit. In the unlikely event that the wireless power transmission is diverted, the power beam will automatically be defocused.

Since 1968, numerous experiments have been conducted to assess how best to transmit power from satellites to Earth's surface with maximum efficiency and minimal environmental impact. Clearly, additional experimentation and testing will be needed as installation and implementation nears.

In NASA's 1997 "Fresh Look at Space Solar Power," the analysis team wrote, "Although systems-level validation of key technologies, such as power conversion and large-scale wireless power transmission (WPT) have not occurred, component-level progress has been great." (Mankins 1997, p. 8) As the first Sunsat start-ups are launched, human safety and environmental protection will be primary concerns for successful implementation.

#9: Why Is Space Solar Power Important for Nations?

Answer: Energy from space has reached a point of priority for three strategic reasons:

1. All known energy supplies will be insufficient to keep up with projected worldwide demand. The U. S. Energy Information Administration forecasts that total world consumption of marketed energy will have increased from 2007 to 2035 by 49%. This means that energy use will have nearly doubled in less than 30 years (U. S. Energy-World Energy 2010).
2. About 80% of the current energy supply is in the form of fossil fuels. Greater diversification of energy sources will be required, with the long-term goal being to find energy sources that are affordable, clean, renewable and available to everyone. Of pressing importance is the need to break the bonds that link electricity production to tons of coal and that bind transportation to barrels of oil.

3. Guaranteed access to non-polluting energy is a controlling variable for local and regional security, economic and social development and a good quality of life. Health care, transportation, telecommunications, education, heating, lighting, refrigeration, food production, and water purification are among the basic necessities of modern civilization that depend on a ready and a reliable source of energy.

Electricity is one of the most flexible, cost effective and non-polluting sources of power at the point of use, and energy from space will be key to universal access to this form of power.

Looking to the future, having one's own space power grid will assure that baseload electrical power is available 24/7 and can be distributed to all users everywhere. Eventually, electrical power will be exchanged among countries using the power grid that is above all countries. Were China, Japan and India, for example, able to construct their own electrical power grids in space, contributing to an all-Asia space solar power cooperative, these three nations could better assure that all segments of their societies would benefit, and that all countries within the region could more equitably participate in the economic stimulus that energy from space would provide.

Were similar space-based solar systems implemented by the European Union, the EU's solar satellite beams could also target Africa and the Middle East. Likewise, a North American constellation could reach to the Caribbean, Central America and all of South America. And, under such conditions, reversing directions from south to north would be both plausible and very likely.

As a special bonus, each of the participating countries can expect substantial reductions in the huge transport and environmental costs associated with coal and petroleum importation.

#10: What Is the Next Step for Solar Power Satellites?

Answer: From the perspective of the year 2012, the variance between what is likely to happen and what should happen is huge. The likely approach will be that Earth citizens will just continue day-to-day doing whatever they are currently doing to "make a living," without thinking too much about the long-term consequences of their lifestyle and business decisions on future generations.

For those who have a sense of social responsibility and do not feel powerless to change the course of human events, there are things we can actually do to address our energy predicament. These are informing ourselves about the facts of the current human predicament. Recommended reading is:

- Prince Charles' address on "The Future of Food," in which he says, "In a global ecosystem that is, to say the least, under stress, our unbridled demands for energy, land and water puts overwhelming pressure on our food systems. I am not alone in thinking that the current model is simply not durable in the long term" (Land Report 2011), and
- The collection of essays entitled "The Rights of Nature: The Case for a Universal Declaration of the Rights of Mother Earth," which seeks to "pursue human well being in a manner that enhances and maintains the integrity, balance and health

of Mother Earth instead of undermining it" and encourages "the peoples and nations of the Earth to work together to replace the exploitative values, world views and political, economic and legal systems with those that respect and defend the rights and harmonious co-existence of all beings" (Council of Canadians 2011).

Certainly, we can be personally engaged in conserving, recycling and squeezing maximum efficiency from the energy resources we have. We can also aid in the search for alternative approaches, one of which is to draw more directly on the clean and abundant power source that the sun makes available to all society.

Whether on a collaborative, competitive or some other basis, some of us should be at work designing the next-generation satellites that will be needed to sustain a high quality of life on Earth. We should be putting the space solar power option on the table for public discussion. We should be looking at plausible solar power satellite applications and launching demonstration projects.

Is it implausible that some of the estimated $1 to $3 trillion in private equity that is sitting on the sidelines in the United States as "dry powder" will find a comfortable home as capital investments in such energy alternatives as space solar power? Is it out of the question that international capital markets, whether institutional or private, can be tapped as a source of "impact investing" in which for-profit investments provide solutions to social and environmental challenges? If not, now is the time for this capital to be put to work helping to lift our global economies out of impending decline.

To the extent that companies and investors perceive that they have a public responsibility beyond their responsibility to maximize financial returns, it is conceivable that strategic investing that advances space-based solar power development need not come only from taxpayer funding. Incubating innovation and mobilizing attention to such game-changing solutions as new energy development through private sector capital markets makes sense when those investments hold commercial as well as social and environmental promise (Bugg-Levine 2011).

It would be helpful were new energy policies negotiated to undergird this virtuous chain of events. Ralph Nansen's five criteria by which any new energy source should be judged seem like a sensible place to start. Nansen says such energy should be: (1) a non-depletable, sustainable resource; (2) non-polluting, environmentally clean; (3) low-cost, over a long period of time; (4) in usable form, and (5) be available to all (Nansen 1995, pp. 6–7).

The framing of new energy policy—community by community, country by country—will automatically open the door to discussion about energy alternatives. The hard data that prioritizes alternative energy as a way of tackling the forces driving climate change and economic stagnation will assure that solar power satellites are on the list. Wherever new energy policy is taken seriously, space goals, strategies, objectives and tactics are more likely to follow and proof-of-concept projects are more likely to gain funding, with significant contributions from the private sector as well as from government.

This author takes the position that once energy from space gains greater visibility as an option among policymakers, the press will cover such deliberations, the public will become better informed, and the younger generation will readily see and promote its benefits.

References

Ayma, E. M., et al. 2011. *The Rights of Nature: The Case for a Universal Declaration of the Rights of Mother Earth*. The Council of Canadians (Ottawa), Fundacion Pachamama (Quito) and Global Exchange (San Francisco).

Belvin, W. K., J. T. Dorsey, & J. J. Watson. 2010. Solar power satellite development: Advances in modularity and mechanical systems. *Online Journal of Space Communication*. http://spacejournal.ohio.edu/issue16/belvin.html. Accessed 1 May 2011. This article was adapted from a paper delivered to the International Symposium on Solar Energy from Space, Toronto, Ontario, Canada, September 8–10, 2009.

Bugg-Levine, A. 2011. Impact investing: Harnessing capital markets to drive development at scale. Rockefeller Foundation in New York. http://beyondprofit.com/. Accessed 19 July 2011.

Gopalaswami, R. 2010. Sustaining India's economic growth. *Online Journal of Space Communication*. http://spacejournal.ohio.edu/issue16/gopal.html. Accessed 20 May 2011.

Howell, J. T., M. J. O'Neill, & J. C. Mankins. 2006. High voltage array ground test for direct drive solar electric propulsion. *Acta Astronautica 59*/1-5: 206–215. This paper was adapted from a presentation given at the 56th International Astronautical Congress, Fukuoka, Japan. October 17–21, 2005.

Jacobson, M. Z. & M. A. Delucchi. 2009. A path to sustainable energy by 2030. *Scientific American 301*(5): 58–65.

Ji, G., H. Xinbin, & W. Li. 2010. Solar power satellites research in China. *Online Journal of Space Communication*. http://spacejournal. ohio.edu/issue16/ji.html. Accessed 15 June 2011.

Klotz, I. 2010. NASA finds cause of *Voyager* 2 glitch. *Discovery News*, May 18. http://news.discovery.com/space/nasa_finds_cause_of_voyager_glitch.html. Accessed 1 July 2011.

Kollipara, A. 2009. Economics of space solar power. Presentation given at the International Space Development Conference, Orlando FL, May 2009.

Mankins, J. 1997. A fresh look at space solar power: New architectures, concepts and technologies. 38th International Astronautical Federation, NASA, 1997. IAF 97R.2.o3.

Nansen, R. 1995. *Sun power: The global solution for the coming energy crisis*. Seattle WA: Ocean Press.

National Space Society. 2011. www.nss.org/settlement/nasa/Contest/. Accessed 1 September 2011. A revised version with technical brief appears at www.spacejournal.org/issue17.

Porter, A. & C. Clover, Ten UK nuclear power stations by 2020. *The Telegraph*, 10 January 2008. http://www.telegraph.co.uk/.

Poynter, M. & A. L. Lane. 1984. *Voyager: The Story of a Space Mission*. New York: Atheneum.

Preble, D. 2011. Answers to space solar power questions. Personal communication with the author, 3 March 2011.

Prince Charles. 2011. The Future of Food. The Land Institute. Salina, Kansas, Summer 2011.

Sato, S. & Y. Okada. 2009. Mitsubishi, IHI to join $21 billion space solar project. *Bloomberg*. http://www.bloomberg.com. Accessed 15 September 2009.

Strickland, J. Space solar vs. base load ground solar and wind power. *Online Journal of Space Communication*. Winter 2010. www.spacejournal.org.

Strickland, J. 2011. Answers to space solar power questions. Personal communication with the author, March 1, 2011.

U. S. Energy Information Administration. 2010a. Updated capital cost estimates for electricity generation plants. http://www.eia.gov/oiaf/beck_plantcosts/. Accessed 1 July 2011.

U. S. Energy Information Administration. 2010b. World energy demand and economic outlook. http://www.eia.gov/oiaf/ieo/world.html. Accessed 18 June 2011.

Werbos, P. 2011. Answers to space solar power questions. Personal communication with the author, March 1, 2011.

World Summit on the Information Society. 2003. Declaration of principles. http://www.itu.int/wsis/docs/geneva/official/dop.html. Accessed 10 June 2011.

Glossary

Antenna The technological devices used for transmitting and receiving energy via radio frequency or light waves.

Apogee The point at which an elliptically orbiting solar satellite is most distant from the surface of Earth.

Attenuation The loss of power in transmission of a wireless energy beam from transmitter to receiver.

Co-location The positioning of more than one satellite at the same (approximate) location made possible by differentiating their electromagnetic frequency assignments.

Downlink The transmission path of an energy beam traveling from a solar power satellite to its rectenna (receiving antenna) on the ground.

Electromagnetic frequency The wavelength (or cycles per second) of a particular radio wave measured in hertz (Hz), where 1 kHz = 1,000 cycles per second; 1 MHz = 1,000 kHz; 1 GHz = 1,000 MHz, and 1 thz = 1,000 GHz.

Electromagnetic spectrum This is a means of designating and assigning the full range of electromagnetic radiation possible using radio and light waves.

FCC The U. S. Federal Communications Commission, which has responsibility for domestic regulation of space communication, frequencies and related matters.

Footprint The area on the ground where energy beams transmitted from space can be received.

GEO Geosynchronous Earth orbit is the unique location where a solar power satellite will move at the same speed as Earth's rotation and is located some 36,000 km above Earth's equator.

GTO The geostationary transfer orbit is a location in space where a solar power satellite can be placed near Earth just prior its being moved into a higher GEO orbit.

Ground segment The terrestrial components of a solar power satellite or satellite system.

D.M. Flournoy, *Solar Power Satellites*, SpringerBriefs in Space Development, DOI 10.1007/978-1-4614-2000-2, © Don M. Flournoy 2012

Intelsat From 1964 to 2001, the International Telecommunications Satellite organization was a non-profit intergovernmental consortium providing global communications services. It now is the world's largest commercial operator of communication satellites with 52 satellites in its fleet.

Inter-satellite links These are wireless radio, optical or energy transmission from space satellite to space satellite.

ITU The International Telecommunications Union is a regulatory agency of the United Nations based in Geneva, Switzerland.

LEO The low Earth orbits that operate at approximate altitudes of 160 to 2,000 km above Earth.

MEO The medium Earth orbits, sometimes called intermediate Earth orbits, that operate at approximate altitudes of 8,000 to 20,000 km from earth.

Nanosats These are micro-satellites that weigh less than 10 kg (22 lbs).

On-board processing Digital intelligence designed into the components of the satellite bus to make solar power satellites more than passive in-space platforms.

Perigee The location in an elliptical orbit in which the solar satellite is nearest Earth.

Pilot signal This is a microwave beam transmitted from the center of a rectenna (rectifying antenna) on the ground to the power transmitting antenna on a solar power satellite to assure continuous control over its energy beam.

Radiant energy This is emitted or received electromagnetic power that travels in space as a wavelike motion similar to radio and light waves.

Rectenna This is an Earth receiving antenna for electrical power produced by a solar power satellite.

Signal interference This is out-of-band and incompatible frequencies, unwanted electrical signals or noise causing degradation of electromagnetic signal reception.

Solar power satellite This is a platform for gathering the Sun's energy in space and transmitting it to Earth, sometimes referred to as a Sunsat or powersat.

Space segment This is the in-space infrastructure, component parts and operations of solar power satellites, which includes the propulsion and robotic systems that move material around in space.

Spot beam This is an energy transmission from a solar satellite that is focused on a designated Earth location.

WARC The World Administrative Radio Conference is a bi-annual meeting hosted by the ITU to discuss issues affecting radio frequency allocation and use.

About the Author

Don M. Flournoy, Ph.D., is a professor of telecommunications at the Scripps College of Communication at Ohio University, Athens, Ohio. He is the founding editor of the *Online Journal of Space Communication* in publication since 2002. An active member of the National Space Society, Prof. Flournoy served for two terms (2002–2008) on the Board of Directors of the Society of Satellite Professionals International.

Prof. Flournoy took his Ph.D. from the University of Texas, and earned a post-graduate Associateship from the University of London. He was Assistant Dean, Case Institute of Technology, Cleveland OH (1965–1969); Associate Dean, State University of New York/Buffalo NY (1969–1971); and Dean of the University College, Ohio University (1971–1981). For 15 years, Dr. Flournoy was director of a research center, the Ohio University Institute for Telecommunication Studies. He is the author of eight books and numerous articles, including writings on space.

Index

A

Agriculture, 12–13, 19, 44, 73–74
Albrecht, M., 77
Asimov, I., 87

B

Belvin, W.K., 24, 25
Betancourt, K., 57–60, 62, 65
Bienhoff, D., 4, 36, 37
Biomass, 12, 15, 90–91
Boeing Company, 9, 13, 25, 30, 47

C

California, 14–15, 21, 25, 52, 64, 82, 92
CAST. *See* China Aerospace Science and Technology Corp.
Chapman, P., 33, 34
China
    aerospace technology, 70
    economic development, 70
    energy, 6, 52, 69, 70, 74, 89
    greenhouse gas emissions, 69
    SPS demonstration, 71
China Aerospace Science and Technology Corp. (CAST), 36, 56, 57, 67–68, 70–71
Clarke, A.C., 87
Climate change, 7, 10–11, 14, 23, 50, 51, 73, 74, 98
Communication satellite (Comsat), 1–6, 26, 29, 31, 45, 53, 55, 61, 63–66, 84, 91–92, 95

D

Davis, D.E., 2, 13, 16, 25
Desalination, 14, 52, 82
Dessanti, B., 39–41
Disaster relief
    recovery, 16

E

Elbert, B., 32
Electricity
    baseload, 20–21, 93
    utilities, 12
Electromagnetic spectrum, 6, 45, 63
Energy
    alternative, 7, 9, 41, 50, 90, 91, 98
    geothermal, 9, 15, 50–51, 90
    hydroelectric, 91
    renewable, 9, 15, 19, 23, 40, 50, 64–65, 69, 70, 72, 96
    supply, 8, 11, 13, 14, 34, 39, 41, 96
Environment protection, 69
Export control
    licensing, 56

F

Fanpei, L., 56
Flournoy, D.M., 48–50, 56, 67

G

Garretson, P., 75
Garver, L., 35

Georgia Institute of Technology, 39, 40
Geosynchronous orbit (GEO)
      (satellite), 23, 32, 63, 83
Gibbons, J., 41–45
Glaser, P., 20, 41, 87, 96
Global warming, 10, 11, 51
Globus, A., 52, 53
Gopalaswami, R., 72–74, 90

H
Hopkins, M., 35
Hsu, F., 10, 11, 19–21, 65

I
India
      economic growth, 72–74
      energy policy, 73
      launch vehicle, 72
      rural electrification, 52, 74
India Space Research Organization (ISRO), 72
Indo-U.S. partnership, 50–51
International satellite (Intelsat), 6, 83, 91, 92
International Space Station (ISS), 2–4, 33,
      77, 87
International Telecommunications Union
      (ITU)
      millennium development goals, 55
      orbit, 63, 82–83
      spectrum, 63
ISRO. See India Space Research Organization
ISS. See International Space Station
ITAR
      international trade, 56
      munitions, 56
ITU. See International Telecommunications
      Union

J
Japan
      energy from space, 89
      wireless power transmission, 7
JAXA (Japanese Space Agency), 75, 89
Ji, G., 89
Johnson, N., 83

K
Kalam, A., 49–50, 72, 74, 75
Kaya, N., 76
Kollipara, A., 90
Komerath, N., 40

L
Landis, G.A., 13, 14, 23
Launch vehicle (LV)
      booster, 31
      propellant, 36
      provider, 29, 30
LEO. See Low Earth orbit
Liability Convention, 57–59
Low Earth orbit (LEO), 4, 22–23, 25, 29,
      32–34, 36, 37, 45, 46, 56, 60, 70–72,
      76, 77, 87, 89, 92–94
LV. See Launch vehicle

M
Mankins, J.C., 12, 21–23, 76,
      93, 96
Mars rover
      Voyager, 80
MEO. See Middle Earth orbit
Middle Earth orbit (MEO), 22–23, 46
Musk, E., 35

N
Nansen, R.H., 30, 31, 47, 48, 98
NASA
      space shuttle, 31
National Space Society (NSS), 10, 16, 34, 35,
      47–50, 62, 74, 88, 93
NSS. See National Space Society
Nuclear energy, 7, 50, 90

O
Obama, B., 48, 49, 56
Oberst, G., 55
Ohio
      GRID Lab, 15
      Ohio University, 3, 15, 16, 82
Online Journal of Space Communication
      Space Journal, 3, 10, 14, 16, 19, 30, 33, 48,
         57, 62, 68, 72, 91
Orbital position (slot), 63
Outer Space Treaty, 57–58

P
Photovoltaic (PV)
      arrays, 13, 80, 95
      cells, 2–3, 19, 26, 36, 95
Potter, S.M., 4
Power beam
      density, 5, 25, 43, 62

Powersat (Sunsat), 1–17, 19–26, 29–37,
    39–48, 53, 57, 59, 65, 67–78, 91–92, 96
Preble, D., 6, 93, 94
Public health and safety, 62

R
Rajagopalan, R.P., 50, 51
Rectenna
    ground receiver, 5
    rectifying antenna, 20
Registration Convention, 57, 59–60
Research and development, 7, 23, 67, 68,
    70–72
Reusable launch vehicle (RLV), 4, 21, 31, 34,
    36, 72, 74
RLV. *See* Reusable launch vehicle
Robotic technology (robotic assembly), 21, 37

S
SBSP. *See* Space-based solar power
SIG. *See* Space Island Group
Society of Satellite Professionals International
    (SSPI), 16, 53, 82
Solar
    concentrators, 8
    power, 1–10, 12–17, 21, 23, 29, 33, 37,
        39, 40, 47, 49–53, 55, 62, 69–76, 80,
        87–91, 93–95
    storms, 83–84
Solaren, 8, 14–15, 64
Solar power satellite (SPS), 1–10, 12–14, 16,
    17, 20–26, 29–33, 41–53, 57–60, 62,
    65–77, 79–85, 88–99
Space assembly
    business, 6, 7, 11, 13, 53, 84
    collisions, 59–60, 82–83
    communication, 2–6, 26, 29, 45, 61, 84
    energy, 1–6, 9–10, 13, 15, 19, 21, 29–31,
        37, 41, 52, 81
    launch, 3–5, 19, 29–32, 37, 58, 79–80
Space-based solar power (SBSP), 4, 7, 17, 19,
    21, 34, 50, 51, 57, 59, 62–65, 69, 74,
    75, 77, 79, 81, 93, 98
Space debris, 60, 83

Space Energy Group, 19, 21, 51, 62, 65
Space Island Group (SIG), 21
Space power grid (SPG), 39, 40, 97
Space solar power (SSP), 3, 6, 7, 8, 10,
    12–16, 20–25, 30, 33, 39–41, 48–50,
    52–53, 62, 67, 68, 70–76, 79–82, 87–99
Space Solar Power Workshop, 22
SpaceX, 32, 35, 36
SPG. *See* Space power grid
Spot beam, 3, 46, 62
SPS. *See* Solar power satellite
SSP. *See* Space solar power
SSPI. *See* Society of Satellite Professionals
    International
Strickland, J., 91, 93
Sunsat (solar power satellite), 1–17, 19–26,
    29–37, 39–48, 59, 65, 67–78, 91–92
Sun synchronous
    orbit, 16, 41
    satellite, 17, 41

T
Tesla, N., 20, 96
Tobiska, K., 14
Turning Point Solar, 13

V
Voyager, 92, 93

W
Wallach, M.I., 62–65
Werbos, P., 84, 93, 94
Wireless power transmission (WPT)
    laser, 71
    microwave, 71
    mirror, 16, 43
    signal interference, 84
Woodcock, G., 9, 10, 34, 35
WPT. *See* Wireless power transmission

Z
Zacharilla, L., 53